FROM MELANCHOLIA TO PROZAC

FROM MELANCHOLIA TO PROZAC

A History of Depression

Clark Lawlor

OXFORD
UNIVERSITY PRESS

OXFORD
UNIVERSITY PRESS

Great Clarendon Street, Oxford OX2 6DP

Oxford University Press is a department of the University of Oxford.
It furthers the University's objective of excellence in research, scholarship,
and education by publishing worldwide in

Oxford New York

Auckland Cape Town Dar es Salaam Hong Kong Karachi
Kuala Lumpur Madrid Melbourne Mexico City Nairobi
New Delhi Shanghai Taipei Toronto

With offices in

Argentina Austria Brazil Chile Czech Republic France Greece
Guatemala Hungary Italy Japan Poland Portugal Singapore
South Korea Switzerland Thailand Turkey Ukraine Vietnam

Oxford is a registered trade mark of Oxford University Press
in the UK and in certain other countries

Published in the United States
by Oxford University Press Inc., New York

British Library Cataloguing in Publication Data
Data available

Library of Congress Cataloging in Publication Data
Data available

Typeset by SPI Publisher Services, Pondicherry, India
Printed in Great Britain
on acid-free paper by
Clays Ltd, St Ives plc

ISBN 978–0–19–958579–3

1 3 5 7 9 10 8 6 4 2

CONTENTS

ACKNOWLEDGEMENTS

I apologise sincerely to anyone I have forgotten to thank in the following acknowledgements: making a book is a complicated and long-term affair, and my fading grey-matter has no doubt let me down at some point. My main thanks are to Helen and Bill Bynum, without whose initiative this book would not exist. Helen has been a marvellous editor and has frequently gone beyond the call of duty in responding to my pestering – I have also greatly enjoyed our wildlife discussions as a sideline! The anonymous readers and Delegates at OUP were tremendously helpful in shaping the book and I wish I could thank them personally. I am also very grateful to the team at OUP, who have patiently shepherded me through the production process: Latha Menon, Emma Marchant (who has borne the brunt of my questions), Eleanor Ashfield, Fiona Vlemmiks, Mike Protheroe and Kevin Doherty. All have been meticulous in their attention to this project. Alasdair McCartney at the Wellcome Library has expedited the images with great efficiency.

I have been fortunate to have participated in the *Before Depression* project at Northumbria University, generously funded by the Leverhulme Trust and further supported by the School of Arts and Social Sciences. Allan Ingram (the Director of the project) has been a great friend to me throughout my career, and the project team have made a real intellectual community in which I have been very happy. So- many thanks to

Leigh Wetherall Dickson, Richard Terry, Stuart Sim, Charlotte Holden (who has also done most of my index), Pauline Morris and Diane Buie. The many participants in the project beyond Northumbria have informed my thinking, whether consciously or not. My colleagues in the Department of Humanities have been ever-encouraging, with special thanks to Alan Harvey (for tea and Spurs/Argyle diversions!) and David Walker. The University and the School have also given me the research time and money to complete this book.

I have road-tested some of my material on various unsuspecting victims: the Johnson material was patiently endured by the Oxford University Restoration to Reform seminar, and insightful comments provided by Christine Gerrard, Abby Williams, Roger Lonsdale, Ros Ballaster and Freya Johnston. Thanks too to Catherine Batt for inviting me to speak at the highly stimulating Leeds University 'Reading Medical Cultures Across Time' event. At Durham I have been aided by helpful thoughts from Jane Macnaughton, Corinne Saunders and David Fuller, and at Newcastle by Jonathan Andrews and Jenny Richards. Further afield, I am grateful to Akihito Suzuki, Junko Kitanaka, Alan Bewell, Michelle Faubert, Tristanne Connolly, Valerie Maffre and Gert Brieger for support intellectual and practical. George Rousseau introduced me to medical humanities and depression many years ago, and Alison Brabban has provided an expert view on many aspects of the professional issues involved in contemporary depression.

Thanks to my family, especially my parents, Jim and Eileen Lawlor, for putting up with my occasional absence when meeting production deadlines. My parents-in-law, Eliot and Ginette Dayan, have as usual enabled me to get on with work at crucial points.

Finally, I dedicate this book with all my love to my wonderful boys, Max and Mikey, and my equally wonderful wife, Margaret Dayan. Without her moral support and saintly willingness to cover for me despite her own busy and successful career as a neuro-ophthalmologist, I could not have finished this book.

LIST OF ILLUSTRATIONS

PROLOGUE: DR SAMUEL JOHNSON (1709–84)

About this time he was afflicted with a very severe return of the hypochondriack disorder, which was ever lurking about him. He was so ill, as, notwithstanding his remarkable love of company, to be entirely averse to society, the most fatal symptom of that malady. Dr. Adams told me, that as an old friend he was admitted to visit him, and that he found him in a deplorable state, sighing, groaning, talking to himself, and restlessly walking from room to room. He then used this emphatical expression of the misery which he felt: 'I would consent to have a limb amputated to recover my spirits.'[1]

Boswell's *Life of Samuel Johnson*

D r Samuel Johnson, the 'Great Cham' of English Literature, towering giant of the Enlightenment, editor of the major Dictionary in the language, and champion of Reason, may seem an odd figure to pick as a case study for depression, but this quotation as reported by James Boswell, another great depressive (or 'hypochondriack', as they also termed it in those days), gives good reason for that choice. That depression has been fashionable through-out its history is an odd fact, and one that we will constantly encounter in this book, but the glamour of Johnson's literary fame did not save him from the crushing despair and 'misery' characteristic of the condition. Compared to physical pain, the mental turmoil that Johnson endured was far worse.

How did Johnson come to be like this? How did his age view him and his disease? Depression is always different at a personal level, in the sense that each individual has a specific psychologi-cal and social position, and yet the same within a specific social group, in that cultures have their own ways of constructing narratives around a disease: each has its own 'story', or even its own 'biography'. These stories evolve, or perhaps change drasti-cally in a short space of time: the story of depression is one in which we seem to witness a comparatively consistent disease phenomenon that is nevertheless endlessly reconceptualised and lived according to the experience of the particular culture and individual concerned. Johnson's story helps us understand how this might happen.

Johnson's history

'This is my history; like all other histories, a narrative of mis-ery.'[2] Johnson was talking about the physical ailments at the end of his life, gout and asthma among them, but—in his case as in other depressives—it is difficult to disentangle physical symptoms from the psychological malady. It has been a tenet of modern medicine to divide mind and body, but much recent research has been returning to a truth previously and com-monly acknowledged: that mind and body go hand-in-glove (see Fig. 1). Johnson's problems started in his youth, and even before, if one believed the hereditary theories of depression current in his day. His childhood scrofula—which even involved an attempt at cure by a quasi-mystical touching of the sufferer by the monarch to cure the 'king's evil'—had a profound effect on his health, and no doubt contributed towards his depression. His friend, James Boswell, provides us with a striking verbal portrait of Johnson at the very end of his *Life of Samuel Johnson*:

> His figure was large and well formed, and his countenance of the cast of an ancient statue; yet his appearance was rendered strange and somewhat uncouth, by convulsive cramps, by the scars of that distemper [scrofula] which it was once imagined the royal touch could cure, and by a slovenly mode of dress. He had the use only of one eye; yet so much does mind govern and even supply the deficiency of organs, that his visual perceptions, as far as they extended, were uncommonly quick and accurate. So morbid was his temperament, that he never knew the natural joy of a free and vigorous use of his limbs: when he walked, it was like the struggling gait of one in fetters; when he rode, he had no command or direction of his horse, but was carried as if in a balloon.[3]

'Temperament' in this period could mean both physical and mental qualities, and we find the word 'morbid' being applied to both in the case of Johnson. Johnson could be portrayed as a comedy eccentric, and Boswell's description does play up the image to a certain extent, but to concentrate on his physical ailments would be to hide his psychological struggles. Boswell then moves on to Johnson's depression, and suggests the reasons for it: 'He was afflicted with a bodily disease, which made him often restless and fretful; and with a constitutional melancholy, the clouds of which darkened the brightness of his fancy, and gave a gloomy cast to his whole course of thinking.'

Fig. 1 Samuel Johnson—as depicted by physiognomists attempting to capture the downcast facial expressions of the melancholic. Physiognomy ('knowledge of nature') was also a popular form of diagnosing and representing depression. The illustrations of depressive individuals for the works of Johann Caspar Lavater and artistically assisted by Thomas Holloway were attempts to assess character through external, physical features. Physiognomy was a classical concept which had regained status in the seventeenth century. *c.*1789. (*Wellcome Library, London*)

'Melancholy' was another one of the terms for depression in this period and, although the word 'depression' appears frequently in the eighteenth century, especially in works like Boswell's *Life of Samuel Johnson*, it is not our concept of depression, which arose towards the end of the nineteenth century and continues to evolve in the present day. Melancholy, hypochondria, spleen, and vapours were all terms for what we now call depression, and all could be as vague (or as specific) as our present variety of definitions and explanations. When Boswell comments—as he frequently does of Johnson—that 'he laboured under a severe depression of spirits', he is referring not to a specific disease (melancholy) but a physical and psychological state in which the 'spirits' are pressed down, lowered, unable to function properly.[4]

Johnson's father, Michael Johnson, was a primary influence on his melancholy. Boswell points to the hereditary nature of melancholy—as he saw it—in Johnson's case. Michael was

> a man of a large and robust body, and of a strong and active mind; yet, as in the most solid rocks veins of unsound substance are often discovered, there was in him a mixture of that disease, the nature of which eludes the most minute enquiry, though the effects are well known to be a weariness of life, an unconcern about those things which agitate the greater part of mankind, and a general sensation of gloomy wretchedness. From him then his son inherited, with some other qualities, 'a vile melancholy', which in his too strong expression of any disturbance of the mind, 'made him mad all his life, at least not sober'.[5]

Boswell's emphasis here falls on sadness rather than anxiety, although both emotions were fundamental to Johnson's depression. Johnson inherits this 'unsound substance' from his father, a deep-seated depression that renders Johnson—in his own words—'mad all his life'. 'Sober', in this context, means 'rational', 'sane'. Boswell censures Johnson's 'too strong expression of any disturbance of the mind', because Boswell wanted to downplay Johnson's depression as the mild affliction of genius. To Johnson himself, the agonies of depression were much more severe, as all forms of his writing demonstrate.

Guilt, idleness, religion, and mothers

The later apparent paradox of Johnson's tremendous productivity in tandem with his genius is also partly explained by events in his early years. Driving Johnson to his grand achievements, suggests Boswell, was his perfectionism, his sense of guilt about not using his considerable talents to the full: 'the solemn text, "of him to whom much is given, much will be required", seems to have been ever present to his mind, in a rigorous sense, and to have made him dissatisfied with his labours and acts of goodness'.[6] The biblical quotation refers to the religious guilt that also plagued Johnson, his terror of death and judgement, and his constant self-flagellation for his so-called idleness, which was a constant theme in his writings. The idea of God was rammed into Johnson's mind from an early age—at three his mother, Sarah, impressed on her genius of a son the need to avoid hell. His genius made him more capable of internalising the message of an angry, Old Testament deity. Johnson's mother also made him read *The Whole Duty of Man* every Sunday, a religious text that warned of dire punishment in fire and brimstone if the pious Christian did not perform extraordinary labours of abjection in the service of his ever-watchful Maker. One does not need to be Freud to see trouble ahead for Johnson, with the crushing weight of Christian guilt bearing down upon him for the rest of his life.

Johnson later rebelled as a teenager and college student, and told Boswell and the poetess Anna Seward (the 'swan of Lichfield') that 'I myself was for some years totally regardless of religion. It had dropped out of my mind. It was at an early part of my life. Sickness brought it back, and I hope I have never lost it since.' To Boswell's amused comment, '"My dear Sir, what a man must you have been without religion! Why you must have gone on drinking, and swearing, and—" Johnson replied (with a smile) "I drank enough and swore enough, to be sure".'[7] Johnson's return to the fold was also motivated by reading William Law's *A Serious Call to a Devout and Holy Life* in 1729, 'overmatched' in his young hubris by Law's powerful arguments in favour of Christianity. It is hard to imagine Johnson's mighty and rational intellect being overmatched by anything, but his childhood reading was too deep-rooted to be susceptible to rational analysis. It was not enough merely to correct wrong thoughts here, because Johnson was in the grip of profound emotions: sadness and fear.

Johnson's life was long, and his depression came and went with varying degrees of severity and with different forces driving it. Roy Porter found religious guilt to be at the root of all these evils, but a number of different factors fed the 'black dog' of melancholy, as both Boswell and Johnson called it.[8] At Oxford, when his depression first came to the fore (as far as we know), his relative poverty was one cause. His attempt to hide his shame at this lack of money and status no doubt made things worse:

> Dr. Adams told me that Johnson, while he was at Pembroke College, 'was caressed and loved by all about him, was a gay and frolicksome fellow, and passed there the happiest part of

his life'. But this is a striking proof of the fallacy of appear-
ances, and how little any of us know of the real internal state
even of those whom we see most frequently; for the truth is,
that he was then depressed by poverty, and irritated by disease. When I mentioned to him this account as given me by
Dr. Adams, he said, 'Ah, Sir, I was mad and violent. It was bit-
terness which they mistook for frolick. I was miserably poor,
and I thought to fight my way by my literature and my wit; so
I disregarded all power and all authority.'[9]

We can see just how bad Johnson's psychological state
around this time was, not merely in his rebellion against the
college authorities, but in an episode of terrifying 'morbid
melancholy', which occurred 'while he was at Lichfield, in the
college vacation of the year 1729'. Johnson 'felt himself overwhelmed with an horrible hypochondria, with perpetual irritation, fretfulness, and impatience; and with a dejection, gloom,
and despair, which made existence misery. From this dismal
malady he never afterwards was perfectly relieved; and all his
labours, and all his enjoyments, were but temporary interrup-
tions of its baleful influence.'[10] Johnson had 'told Mr Paradise
that he was sometimes so languid and inefficient, that he could
not distinguish the hour upon the town-clock'. There were also
hints of Johnson having suicidal thoughts at this time and in the
period after he prematurely left Pembroke College, Oxford, due
to lack of money. Much later Johnson told Mrs Thrale that she
should 'Get your Children into habits of loving a Book by every
possible means; You do not know but it may one Day save them
from Suicide.'[11]

Boswell's account of this episode also reveals the anticipated
social response to Johnson's condition. Johnson had consulted—
in some desperation—Dr Swinfen, his well-meaning godfather,
and given him an account in Latin of his melancholy. Swinfen

was so impressed by the way Johnson had expressed himself that he circulated the document to a few of his friends. Unfortunately Johnson was highly embarrassed and angry, because Swinfen had 'exposed a complaint of his young friend and patient, which, in the superficial opinion of the generality of mankind, is attended with contempt and disgrace'.[12] As in the twenty-first century, prejudice and scorn would often attend the revelation that a person might be suffering from melancholy, or another 'disorder of the intellect'. No wonder Johnson was never entirely reconciled with Swinfen. At the end of his life, Johnson burnt many of his letters and diaries, one of which may well have been a 'history of my melancholy', probably for similar reasons.

To combat his melancholy, Johnson resorted to a variety of solutions, both physical and psychological. In an early and desperate attempt in 1729 to fight what were clearly distressing, probably obsessive thoughts, Boswell wrote that he 'strove to overcome it by forcible exertions': he 'frequently walked to Birmingham and back again, and tried many other expedients, but all in vain. His expression concerning it to me was "I did not then know how to manage it." '[13] The distance from Lichfield to Birmingham—thirty-two miles—is an indication of the monumental effort Johnson put into fighting the demons that were troubling him in this severe breakdown. His departure without a degree from Oxford, a place in which he put on a show of witty bravado, prompted this crisis of status anxiety. His return to Lichfield had left him with few prospects save stagnation in a cultural backwater. Physical exercise, we know now, releases pleasure-inducing endorphins, but the eighteenth century knew it differently:

> Ease is the utmost that can be hoped from a sedentary and unactive habit; ease, a neutral state between pain and pleasure. The dance of spirits, the bound of vigour, readiness of

enterprise, and defiance of fatigue, are reserved for him that braces his nerves and hardens his fibres, that keeps his limbs pliant with motion, and by frequent exposure fortifies his frame against the common accidents of cold and heat.[14]

Although couched in a different logic of 'nerves', 'spirits', and 'fibres', eighteenth-century medicine mirrored our own in its confidence that keeping the body moving through a regimen of exercise was a good defence against , and even cure for, all kinds of maladies, not merely melancholy.

Oxford and the years after were formative in Johnson's depression: there were approximately two years of very severe crisis and then aftershocks for at least three more. His poem on the fate of 'The Young Author', written when he was 20, was a fair indication of the misery he anticipated and later felt at being cast out into the world with no degree and few prospects of success. His literary works are also unremittingly pessimistic in their assessment of the 'vanity of human wishes' and the pointlessness of human achievement, at least as compared to God's judgement and the possibility of eternal salvation or damnation. Social and psychological factors such as status anxiety, poverty, and religious guilt combined to produce Johnson's depressive episodes.

Having left Oxford, Johnson turned his hand to working as a school usher and schoolmaster while attempting to move into writing for a living. In 1735 he also married 'Tetty', or Elizabeth Porter, a widow twenty years older than himself, and comparatively wealthy with it. Unfortunately the immediate happiness and prosperity that Tetty brought him were to become a further cause of depression in later life. Her death in 1752 was another obvious stimulus to one of the worst flare-ups of his melancholia—a combination of grief and guilt about his neglect of her in the pursuit of his literary career. In 1737 Johnson had left Tetty in his native Lichfield while he and Garrick set off for London. This was a pattern that tended to repeat itself throughout the rest of his marriage, even when he and Tetty had established their home in London. Johnson continually neglected Tetty in order to further his literary career; when it came to Tetty's death, Johnson had an overwhelming burden of guilt to carry.

Boswell relates that Johnson's Preface to the *Dictionary* was surprisingly 'desponding', but that one must take into account 'that miserable dejection of spirits to which he was constitutionally subject, and which was aggravated by the death of his wife two years before. I have heard it ingeniously observed by a lady of rank and elegance, that "his melancholy was then at its meridian".'[15] Johnson's letter to Warton on the subject in December 1754 is a testament to the impact of Tetty's death: 'I have ever since seemed to myself broken off from mankind; a kind of solitary wanderer in the wild of life, without any direction, or fixed point of view: a gloomy gazer on the world to which I have little relation.'[16]

Tetty's death had other consequences: although Katherine Balderston has found 'erotic maladjustment' and a masochistic personality at the root of Johnson's melancholy, such a Freudian reading seems unlikely given the evidence of Johnson's actual writings.[17] What we do know is that Johnson suffered guilt after Tetty's death about his sexual desires, not an uncommon problem for the devout Christian, male or female. He pleaded with God to 'purify my thoughts from pollution' and—most explicitly—found that he was not 'once distracted by the thoughts of any other woman' when in church at Easter 1753.[18] Johnson continued to mourn his wife throughout his life—his guilt about her compounding his pre-existent guilt about idleness, religious devotion, and sexual desire.

In 1737, two years after his marriage to Tetty, Johnson learned of the death of his brother, which may have been a suicide. Nathaniel too may have suffered from depression and a drink problem, and might well have influenced Johnson's own thoughts on suicides. To Johnson, suicides were not mad or 'universally disordered', but so possessed by 'one passion' that 'they yield to it, and commit suicide'.[19] From this time on, however, Johnson started to become successful, however slowly, in his publications. He became a hack writer for the *Gentleman's Magazine*, and published his poem *London* in 1738. He began to work on the *Dictionary* in 1746 and published his poem 'The Vanity of Human Wishes' in 1749. Throughout this time Johnson lurched from one financial problem to another, despite the previous boost of Tetty's money, a fact not calculated to ease Johnson's continuing bouts of depression.

Even the production of the *Dictionary* in 1755 did not bring Johnson peace. Quite the opposite in fact: on finishing his monumental labour, Johnson wrote 'Know Yourself', in which he immediately started fretting about how he was going to occupy his thoughts:

> My task perform'd, and all my labours o'er,
> For me what lot has Fortune now in store?
> The listless will succeeds, that worst disease,
> The rack of indolence, the sluggish ease.
> Care grows on care, and o'er my aching brain
> Black melancholy pours her morbid train. (lines 1–6)

Finding no relief from carousing in the pubs and clubs of London, Johnson seeks new projects: 'I form a grand design;/ Languor succeeds, and all my pow'rs decline' (lines 17–18).

This reads very much like post-partum depression, with the *Dictionary* as Johnson's 'baby'. 'Exhausted, tir'd' (l. 15), Johnson could not manage to throw himself into another sloth- and guilt-dispelling enterprise.

Guilt is often linked to depression, as Johnson himself recognised. His comments in his periodical essays and fiction—often semi-autobiographical—are full of insight into the workings of the depressive mind: 'Life is languished away in the gloom of anxiety, and consumed in collecting resolutions which the next morning dissipates; in forming purposes which we scarcely hope to keep, and reconciling ourselves to our own cowardice by excuses which, while we admit them, we know to be absurd.'[20]

In Johnson's tale of *Rasselas* (1759), the character Imlac lays out the operations of guilt, so clearly applicable to Johnson's case, in detail:

> No disease of the imagination, answered Imlac, is so difficult of cure, as that which is complicated with the dread of guilt: fancy and conscience then act interchangeably upon us, and so often shift their places, that the illusions of one are not distinguished from the dictates of the other. If fancy presents images not moral or religious, the mind drives them away when they give it pain, but when melancholick notions take the form of duty, they lay hold on the faculties without opposition, because we are afraid to exclude or banish them. For this reason the superstitious are often melancholy, and the melancholy almost always superstitious.[21]

'Duty' breeds guilt, and guilt, melancholy—all the more insidious and difficult to ignore than immoral images conjured by the imagination. Johnson's religious superstitions, encouraged by his mother, left him with a legacy of 'dread of guilt': 'Surely I shall not spend my whole life with my own total disapprobation.'[22]

Part of Johnson's guilt complex was his conviction that he was idle, and not doing nearly enough to use his talents according to the demands placed upon him by God. His less famous biographer, Sir John Hawkins, noted in his *Life of Samuel Johnson, LL.D.* the great man's propensity to procrastinate: 'He was, throughout his life, making resolutions to rise at eight, no very early hour, and breaking them. The visits of idle, and some of them worthless persons, were never unwelcome to him' (1781).[23] The struggle with his idleness (it is hard to credit that the astoundingly productive Johnson could have seen himself as lazy) was a continuous theme in the recurrence of his depression. His habit of making resolutions that he then castigated himself for infringing, if not actually breaking, set up a cycle of guilt from which he never escaped. Boswell's report of Johnson's words in middleage on his birthday exemplifies this catch-22: 'I have now spent fifty-five years in resolving; having, from the earliest time almost that I can remember, been forming schemes of a better life. I have done nothing. The need of doing, therefore, is pressing, since the time of doing is short. O GOD, grant me to resolve aright, and to keep my resolutions, for JESUS CHRIST's sake. Amen.'[24]

Despite intervals of happy productivity, Johnson lapsed into crisis from around 1760 to 1767, again largely driven by his religious anxieties. It was lucky indeed that, later in this period, he met Hester and Henry Thrale—this distraction of good company was healthy for Johnson and followed his policy of avoiding solitude. His friendship with the Thrale family proved to be a godsend, as it gave Johnson the context of a healthy family life that he himself had lacked:

> Nothing could be more fortunate for Johnson than this connection [with the Thrales]. He had at Mr. Thrale's all the comforts and even luxuries of life; his melancholy was

diverted, and his irregular habits lessened by association
with an agreeable and well-ordered family. He was treated
with the utmost respect, and even affection. The vivacity of
Mrs. Thrale's literary talk roused him to cheerfulness and
exertion, even when they were alone. But this was not often
the case; for he found here a constant succession of what
gave him the highest enjoyment: the society of the learned,
the witty, and the eminent in every way, who were assembled
in numerous companies, called forth his wonderful powers,
and gratified him with admiration, to which no man could
be insensible.[25]

Mrs Hester Thrale, who later recorded her time with Johnson
in her *Thraliana*, became a psychological support for Johnson,
and biographers have even discussed the idea that they had
some kind of bizarre quasi-sexual relationship that involved
Thrale padlocking Johnson in a certain part of her house and
then handcuffing him. The truth seems to have been that
Johnson wanted some kind of protection from his own feared
madness, and that Thrale was the reluctant provider of a fantasy
of confinement that supplied an external discipline so clearly
lacking in his own psyche. Thrale withdrew gracefully from the
role, recommending Boswell as his 'best physician' and asking
that Johnson 'not quarrel with your governess for not using the
rod enough'.[26]

The management of the mind

By the time Johnson reached his early sixties, the influence of the
Thrales had helped him stabilise his melancholy and, although
he still struggled with his religious guilt, he talked more frequently of his ability to manage his condition. He began to
dispense wisdom to Boswell on how to fight depression, and

was not shy of admonishing Boswell for ignoring his advice to read improving literature: 'I am, I confess, very angry that you manage yourself so ill.'[27] 'Cure' is too strong a word for a condition that Johnson knew he would never fully vanquish. Famously adapting the words of that vital melancholic of the Renaissance, Robert Burton, Johnson told Boswell: 'The great direction which Burton has left to men disordered like you, is this, *Be not solitary; be not idle*: which I would thus modify;—If you are idle, be not solitary; if you are solitary, be not idle.'[28]

Obsessive walking was one means to clear his mind of depressing thoughts, and so was the company of people and any activity that demanded the mind be focused away from the objects that dominated and terrorised it. Distraction was crucial: 'For the black fumes which rise in your mind, I can prescribe nothing but that you disperse them by honest business or innocent pleasure, and by reading, sometimes easy and sometimes serious. Change of place is useful; and I hope that your residence at Auchinleck will have many good effects.'[29] Johnson's travels with Boswell into the Scottish Highlands would serve this purpose of distraction admirably. Johnson had also learned to use psychological techniques more effectively. The mind as well as the body could provide many forms of distraction: 'Talking of constitutional melancholy, he observed, "A man so afflicted, Sir, must divert distressing thoughts, and not combat with them." Boswell: "May not he think them down, Sir?" Johnson: "No, Sir. To attempt to think them down is madness. He should have a lamp constantly burning in his bed chamber during the night, and if wakefully disturbed, take a book, and read, and compose himself to rest. To have the management of the mind is a great art, and it may be attained in a considerable degree by experience and habitual exercise." Boswell: "Should not he provide amusements for himself? Would it not, for instance, be right for him to take a course of chymistry?" Johnson: "Let him take a course of chymistry, or a course of rope-dancing, or a course of any thing to which he is inclined at the time. Let him contrive to have as many retreats for his mind as he can, as many things to which it can fly from itself."[30]

Johnson himself had recourse to arithmetic when he was fighting depressive thoughts.[31]

This 'flight' from the mind is a recurring theme in Johnson's management of melancholy, and involved a variety of possible directions, as Boswell reported in July 1763:

> He mentioned to me now, for the first time, that he had been distrest by melancholy, and for that reason had been obliged to fly from study and meditation, to the dissipating variety of life. Against melancholy he recommended constant occupation of mind, a great deal of exercise, moderation in eating and drinking, and especially to shun drinking at night. He said melancholy people were apt to fly to intemperance for relief, but that it sunk them much deeper in misery. He observed, that labouring men who work hard, and live sparingly, are seldom or never troubled with low spirits.[32]

In previous years, Johnson had been a notorious drinker and freely admitted that it had been a way of escaping from the pain of his existence. Here Johnson prescribed, in the manner of his time, a more holistic approach to the cure of melancholy. Not only was mental distraction needed, but also attention to the management of the six 'non-naturals', or those parts of bodily life that were within the power of men to control, such as our diet, the amount we exercise, the hours we choose to rise and go to bed. Johnson here also taps into the part-myth that the labouring classes were less subject to melancholy: it is likely that people in manual jobs still became depressed, but for different reasons (like poverty) than the sedentary or the scholarly of the higher ranks. One had to be careful, though, as with the great melancholic scholar of the Renaissance, Burton, too much study and reading could have serious consequences: 'Study requires solitude, and solitude is a state dangerous to those who are too much accustomed to sink into themselves.'[33]

It is clear that much of Johnson's depression was fuelled by religious guilt, initially in the form of regret that he had talked against religion in his earlier years, and then because he applied impossible standards of behaviour on himself when older. But what was a cause of his depression was also a means to alleviate its effects. In an essay in the ironically named *Idler*, given Johnson's prodigious productivity, he rejected both pagan philosophies of the Epicurians and the Stoics, and instead pointed to Christianity as the only permanent solution to loss and sorrow:

> Real alleviation of the loss of friends, and rational tranquillity in the prospect of our own dissolution, can be received only by the promises of Him in whose hands are life and death, ... Philosophy may infuse stubbornness, but Religion only can give patience.[34]

It must be remembered, however, that Johnson had a terror of death—increasing with age—and found it difficult to believe that any rational person would not be afraid of their final judgement. Boswell relates an episode of Johnson's religious melancholy from 1777: 'it appears from his *Prayers and Meditations*, that Johnson suffered much from a state of mind "unsettled and perplexed", and from that constitutional gloom, which, together with his extreme humility and anxiety with regard to his religious state, made him contemplate himself through too dark and unfavourable a medium'.[35] Boswell goes on to cite Johnson's own revealing words on his state of mind at the time: 'When I survey my past life, I discover nothing but a barren waste of time, with some disorders of body, and disturbances of the mind, very near to madness, which I hope He that made me will suffer to extenuate many faults, and excuse many deficiencies.'[36]

His prayer on Easter Day illuminates Johnson's perception of himself as fighting a war against depressive thoughts:

> Almighty and most merciful Father, who seest all our miseries, and knowest all our necessities, look down upon me, and pity me. Defend me from the violent incursion of evil thoughts, ... Have mercy upon me, O GOD, have mercy upon me; years and infirmities oppress me, terrour and anxiety beset me.[37]

It is in Johnson's prayers and his literary writings that we find the most moving expressions of his depression and his attempts to cure it. Religious cure is not usually an option for our secular modern societies, and Johnson's was starting out on the long road towards the rule of reason over religious faith, but, as his case demonstrates, Christianity still had a powerful psychological influence over people—even the great systematiser of the Enlightenment like 'Dictionary' Johnson.

The prayer-cure did sometimes work to hold off symptoms: regarding Johnson's odd habit of muttering to himself, Boswell commented: 'Talking to himself was, indeed, one of his singularities ever since I knew him. I was certain that he was frequently uttering pious ejaculations; for fragments of the Lord's Prayer have been distinctly overheard.'[38] Evidently Johnson had

The Worshipfull
Clapham in Surrey
Comissioners of
James the 2 and
affaires of his Ma.

William Hewer of
Essex one of the
the Navy, to King
Tresurer for the
Late Garrison of Tangi..

(84) David playes on the Harpe.

1. Samuel 16.

G Freman delin:

J. Xij. Sculp:

Fig. 2 David playing the harp for a melancholy Saul to ease his mind: this was a famous biblical scene that recurs throughout the literature on melancholy. (*Wellcome Library, London*)

recourse to the incantatory consolations of the most famous prayer in the language, perhaps as comforted by its familiarity as by its sentiments. Johnson himself made the case explicitly on Easter Day 1777 'at church': 'I was for some time distressed, but at last obtained, I hope from the GOD of Peace, more quiet than I have enjoyed for a long time' (see Fig. 2).[39]

Unfortunately for Johnson, the unequal struggle with religious guilt was to continue all his life. Sir John Hawkins described his own attempts to soothe Johnson's mind in the last years of his life: 'These suggestions [to assuage the fear of death] made little impression on him: he lamented the indolence in which he had spent his life, talked of secret transgressions, and seemed desirous of telling me more to that purpose than I was willing to hear.'[40] The barrage of physical ailments that attended Johnson's final years did not help his state of mind, plagued as he said himself by 'great weakness, and of phantoms that haunted his imagination' (1784). Famously, he quoted lines from Macbeth to his physician, Dr Brocklesby:

> Canst thou not minister to a mind diseas'd;
> Pluck from the memory a rooted sorrow,
> Raze out the written troubles of the brain,
> And with some sweet oblivious antidote,
> Cleanse the full bosom of that perilous stuff,
> Which weighs upon the heart?

To which the Doctor wittily and truthfully replied: '—Therein the patient/Must minister unto himself.'[41] To the end, Johnson displayed more concern with his psychological burdens than his physical ones.

Johnson, the great champion of Reason and modernity, was bound into depression by the superstitions of an older era, caught in the middle of a long historical transition from the religious to the secular. Johnson was also in the middle of a shift from the pre-scientific world of medicine to the modern scientific one with which we live today; a shift embodied by the change—to be charted throughout the course of this book—of melancholy to depression.

I

---∞∞∞---

'POOR WRETCH'

[The nymphs] gathered beside Jason,
And with their hands gently drew the mantle from his
    head.
But he turned his eyes away in the other direction
Out of awe for these deities; for he alone could see them,
And, to him in his grief-stricken state, they spoke soothing
    words:
'Poor wretch, why so stricken by absolute helplessness?'[1]

                                            Apollonius, *Argonautica*

This is Jason of the Argonauts, as depicted by Apollonius
of Rhodes in his neo-epic poem *Argonautica*, the tale of
the *Argo*'s voyage. When his ship is stuck on the Libyan
coast and nymphs provide assistance, Jason is paralysed by what
we might now call depression, although it is clear that this is a
depression very much with a cause—after all, he is shipwrecked
on the way home. Even so, Jason's reaction seems excessive for
an Homeric hero. Can what is depicted in a fictional character
by Apollonius be equated with our modern form of depression?
Some think so, and others feel the divide between the two cultures to be so great as to be unbridgeable: the conflict between

Fig. 3 Orestes: is this a depiction of a depressed person in ancient Greek culture? (fourth-century BC red-figure Apulian vase from the Louvre) (*Louvre, Paris, France/Peter Willi/The Bridgeman Art Library*)

these two perspectives on depression as biologically hard-wired or entirely culture bound will recur throughout this book.

Bellerophon is the first Greek melancholy hero, and is depicted in Homer's *Iliad* as 'a lonely wanderer on the Aleian plain, eating out his heart and shunning the paths of men', having offended the gods.[2] We saw Dr Johnson consciously echoing this passage in reference to himself after the death of Tetty. It has been argued that depression has existed from classical times, and that the character of Orestes, in Aeschylus's tragic trilogy, the *Oresteia*, is proof. Orestes, as depicted on a fourth-century BC red-figure Apulian vase (now in the Louvre), is undergoing a rite of religious purification in order to rid him of the murder of his mother, Clytaemnestra, who had been party to the murder of his father, Agamemnon (see Fig. 3).[3] Orestes's depression is manifested in his posture, the downcast eyes and drooping

24

body, drained of all energy. This depression masks his inner turmoil, and it is true that in our own time depression is often interpreted as hiding psychic conflict (not that the concept of the Freudian unconscious was available to the Ancients even if Freud did cull his most famous 'complex'—that of Oedipus— from Greek drama). To consider the issue of ancient depression properly, let us begin with an examination of the highly influential contexts for depression in the classical period.

Definitions—popular versus technical usage

Depression was not called depression in classical writings, either medical or popular. Rather, the term 'melancholia' commonly indicated a long-running mental illness with core symptoms of causeless sadness and fear that derived from an excess of black bile (*melaina chole*), the melancholy humour (one of the four humours supposed to constitute the body). Writers from the Greek school of Hippocrates, which originated in the fifth century BC, considered that 'fear or depression that is prolonged means melancholia'.[4] One complication of this definition is that 'melancholia' in classical literature could involve aggressive madness. The best description of what might mostly closely correspond to modern depression in classical literature—Seneca's *de tranquillitate animi* [i.15–17]— itself observes that there is no name for this condition. Depression existed, but not in much of the violent 'melancholia' named by the literary authors.[5] Fear and sadness were core features of melancholia, but many other symptoms could attend the classical definition of this condition. Nevertheless, the mythologies and medicine of melancholia from the classical period, however construed at the time, have had an enormous influence throughout Western culture up to the present day.

Not all medical writers were humoralists, but in general, and in terms of later influence, the schema of the humours was satisfyingly logical (although without any true physiological basis as we know now) and mapped onto an entire cosmology that embraced the body, nature and time, and even the motions of the stars in later Renaissance versions of the theory. The four humours corresponded to the Pythagorean elements, and consisted of blood (warm and moist—fire), yellow bile or choler (warm and dry—air), black bile or melancholy (cold and dry—earth), and phlegm (cold and moist—water), which in turn dominated spring, summer, autumn, and winter.

Melancholia was not considered a specific disease entity as we would understand it, but the result of an imbalance in the humours: the more severe the imbalance, the more severe the symptoms of melancholia. Although the illness had a physical basis, and could manifest a wide array of physical as well as psychological symptoms (including hallucinations caused by vapours produced by overheated black bile), it was recognised that powerful emotional stimuli could also cause humoral imbalance and thus melancholia. Like the term 'depression' today, its popular usage could be confused: sometimes it might mean generally unusual or 'mad' behaviour. One core distinction in ancient Greek medicine was the tripartite division of madness into phrenitis (frenzy), mania (raving), and melancholy. Melancholy was chronic, not acute like phrenitis, and not feverish; there was typically no raving as in mania. Manic depression, as construed by post-nineteenth-century definitions, bears no relation to the classical forms of mania, in which mania (insanity and delirium) might emerge from melancholia if the melancholia became particularly severe.

Causeless fear and sorrow, thus, were the foundation of melancholy, sometimes accompanied by hallucinations. In our modern, popular picture of depression we have almost no room for this hallucinatory or visionary aspect of melancholy, yet it is fundamental to the Ancient Greek conceptualisation of the disease because of the humoral theory of black bile (μέλαωα χολή) from which the word melancholy proceeds. μέλαωα χολή or *melaina chole* was translated into Latin as *atra bilis* and into English as *black bile*. It is this hallucinatory aspect that appears to be the basis of the famous association of genius with melancholy, the core quality that has driven the fashionability of melancholy and its cognates throughout the centuries.

The basis for these hallucinations was therefore physical. The Hippocratic school had downplayed the importance of religious causes, although it still acknowledged the role of the gods in man's affairs. Literary works, however, still tended to stress the divine causes of man's mental woes. Later in the second century AD, Galen—physician to the Emperor of Rome—(*c.*131–201) put his considerable intellectual weight behind the Greek, Hippocratic version (*c.*460–370 BC) of the essential symptoms and aetiology (causes) of melancholia: 'it seems correct that *Hippocrates* classified all their [melancholics'] symptoms into two groups: fear and despondency. Because of this despondency patients hate everyone whom they see, are constantly sullen and appear terrified, like children or uneducated adults in deepest darkness. As external darkness renders almost all persons fearful, with the exception of few naturally audacious ones or those who were specially trained, thus the colour of the black humour induces fear when its darkness throws a shadow over the area of thought [in the brain].'[6] This imagery (not merely a representation but a physical reality to the Ancients) persists in descriptions of melancholy and depression down the centuries, although differently conceptualised depending on the medical theory underpinning it. The metaphor of the psychological darkness of depression cuts across time periods as it so accurately represents the feeling of losing one's sense of hope, often combined with a feeling of being lost, or having suffered a loss. The classical 'black sun' of depression connected the metaphorical and literal connections of melancholia with the dark innards of the human body and the black bile said to cause general madness, not merely melancholia.

For the Ancients, there was a physical process that led to this darkness: one of the four humours would predispose a person towards a certain personality and linked humoral type. In the case of the black bile or melancholy, a 'natural melancholy' would result from the predominance of the melancholy humour; an 'unnatural melancholy' could come from an excess of melancholic humours being burnt by heating processes such as overexcitement of the passions, poor diet, or a fever.

Key to the production of humours was the digestive system: the heat of the stomach changed food into chyle in the liver, and thence mysterious and superfine yet still physical 'vital spirits' were transmitted to the heart and brain, with these spirits acting as a connection between body and mind, emotion and reason. From the brain came another set of spirits, the 'animal spirits', which were the vital further refined. These spirits ran around the body and, when functioning properly, ensured that all its parts worked in harmony. Imbalance of the humours might cause the body and hence the vital and animal spirits to become excessively hot or cold, dry or moist, thus disorienting the spirits from their proper paths throughout the body. The humours would then affect the thoughts and emotions of the person in proportion to the degree of disturbance in their balance brought about by the extent of heat, cold, dryness, or moisture in the body. Women tended to be colder and more humid than men, although they found it harder to resist overheating and becoming melancholic when possessed by love because they were also supposedly more irrational than men.

When black bile overheated, the burnt remainder was an ash called 'melancholy adust' or 'atrabilious melancholy'. Because the variety of mixtures of the four humours within the human body was potentially infinite, each individual's constitution or temperament was different, although each of the four humours could provide a predominant personality: the sanguine, the choleric, the phlegmatic, and the melancholic. In general a melancholic person was inward looking, often intellectual in the broadest sense of the term, possibly solitary in tendency. Each person's melancholy was unique to them, however subtly: it was not a single disease entity in the way we might now think of depression caused by a chemical imbalance. There was, however, a point at which a temperamental inclination to be melancholy might tip over into a serious illness. This tension between depression as a crushing illness and the milder forms of its manifestation that we now describe as 'melancholy'—its original severity having been lost in the twentieth century— was already an issue in the classical period.

Black bile was a general troublemaker, more so than the other humours. Galen identified three subtypes of melancholy: in the brain, a local problem; in the blood in general (which darkened the skin); and in the hypochondries—the area just below the ribcage—in which the excess of black bile from digestive disturbance caused 'an atrabilious evaporation [which] produces melancholic symptoms of mind by ascending to the brain like a sooty substance or a smoky vapour'.[7] The flatulence from this third type of melancholy earned it the sobriquet of 'windy melancholy' or 'hypochondriacal melancholy' for many centuries to come. Imbalances of black bile and the melancholy resulting from these had many causes, including a bad lifestyle or regime of health—*dieta* as it was known to the Greeks. Too much 'heavy and dark wine' and 'aged cheeses' might set off the condition, as might too much worrying and lack of sleep.[8] Importantly, melancholy might be stirred by the impact of powerful emotions or thoughts on the humours: one of the most famous literary causes for melancholy was to be love, or more precisely the frustration of desire in the pining lover.

Depressive love melancholy—as opposed to the manic form, the heroic, violent, and burning love of Medea, for example— was comparatively rare in classical literature, with the earliest, 'unambiguous' examples from Theocritus, and the tradition of the depressive love melancholic only really getting under way late in the classical period. The lovesick Charicleia in Heliodorus's *Aethiopica*, for instance, falls in love with Theagenes and, not being in possession of her love object, is cast into depressive melancholy. The wise physician Arcesinus has the answer:

> Can you not see that her condition is of the soul and that the illness is clearly love? Can you not see the dark rings under her eyes, how restless is her gaze, and how pale is her face, although she does not complain of internal pain? Can you not see that her concentration wanders, that she says the first things that come into her head, that she is suffering from an unaccountable insomnia and has suddenly lost her

self-confidence? Charicles, you must search for the man to cure her, the only one, the man she loves.[9]

Happily, the lovers are united and a cure effected. Not all these stories ended so well. Medical writers like Aretaeus and Galen had described the role of love in melancholic men and women. These depressive depictions of melancholy love would become the dominant ones in later Western literature, beginning with the medieval period.

The emotions (or 'passions of the soul') were one of the six 'non-naturals', or those factors considered to be within the control of the person; the others were diet, climate, exercise, sleep, and evacuation and retention of bodily substances (not only faeces and urine, although these featured largely). As melancholy was cold and dry, climates, times of the year (the fall), and times of life (middle and old age) that corresponded to these qualities would also predispose people to the illness. Cure of melancholy would consist of attempts to right the humoral imbalance by purging the excess melancholy humour from the blood (bloodletting, leeching, for example) or by certain drugs such as hellebore, effectively a poison that caused immediate diarrhoea and vomiting.[10]

In women, where the retention of the menses was always a danger to health because it meant a build-up of excess humours, attempts might be made to encourage the resumption of such flows from the body. In the case of women, it was thought that marriage (and sex) could cure them of build-ups of harmful blood. This peculiarity of women's unruly physiology has remained a theme throughout Western conceptions of depression and melancholia up to the present day. The same went for a general elimination of the black bile in both sexes, in which

Galen recommended 'exercise, massage and all kinds of active motion' to keep the humours flowing. A diet of warm and moist foods might also counteract the cold and dry melancholy humour.

Aristotle and genius—popular myths of melancholy and depression

Classical medical thought has been fundamental to popular myths about depression: the Greek Hippocratic school paved the way for modern modes of thinking about mental illness by positing natural causes for depression rather than the supernatural. Later, of course, Christianity would blame the devil and/or God's anger for individual suffering. There was a divergence, initially, between the popular, literary, conception of melancholia and the medical one in the classical period. For medics, the influence of the black bile made people depressive, while in literary representation the melancholic was mad, manic, and often violent, most commonly sent by the gods rather than from a physical cause: satirically so in the case of comedy, and heroically so in the case of epic and tragedy.

The nonmedical tradition lacked interest in depressive melancholia because it did not possess the conceptual tools to articulate depression in representation: for the arrival of narrative representations of depression, classical literature required greater interest in interiority, psychological depth, and emotional subtlety. In Greek literature, there was at first a focus on the individual's relationship with society and surface matters.

If literature was slow to realise the potential for intellectual and creative depth in the character of the melancholic, it made up for it later, not least due to the powerful intervention of Aristotle or—more likely—one of his followers who linked melancholy and genius. In the *Problemata* (*Problems*, *c*.350 BC) 'pseudo-Aristotle' stated that men of great ability were melancholics. In these men the rule of the black bile was not diseased, but beneficial, because it was not in excess. Earlier notions of madness as inspired by the divine, which derived from Plato, became secularised in the attribution of genius to the black bile. Such a link went against the later alternative Galenic formulation of the sluggish and dull melancholic, a depressive overwhelmed by the darkness of the black bile.

The 'Aristotelian' position was famously trumpeted in *Problemata* 30:

> Why is it that all men who have become outstanding in philosophy, statesmanship, poetry or the arts are melancholic, and some to such an extent as they are infected by the diseases arising from black bile, as the story of Heracles among the heroes tells? For Heracles seems to have been of this character, so that the ancients called the disease of epilepsy the 'Sacred disease' after him. This is proved by his frenzy towards his children and the eruption of sores which occurred before his disappearance on Mount Oeta; for this is a common affection among those who suffer from black bile.[11]

The Aristotelian author, in a synthesis of the humoral and manic discourses of melancholy, regards the more extreme forms of melancholy as involving frenzy. This included the introduction of epilepsy—another effect of the black bile—in an allusion to the Platonic idea of divine inspiration flowing through the epileptic. Here we are reminded that, although our modern focus is on the scientific conception of the body and disease, the Ancients also believed that divine forces were at play in madness and human life in general. Pseudo-Aristotle's linkage of epilepsy with black bile is a materialist version of the Platonic theory of the frenzied, visionary, creative personality. Pagan Sibyls and soothsayers would be discarded by the melancholy of subsequent times, but the 'enthusiasm' of zealots in Christian sects would be attributed to a form of religious melancholy.

Pseudo-Aristotle forges a strong link between literature, genius, and melancholy in *Problemata* 30: Lysander, Ajax, Bellerophon, Empedocles, Plato, Socrates—all are heroically melancholic. Here Plato's concept of the divine *furor* or crazed inspiration becomes associated with the melancholic hero, who originally was punished by the gods with madness, and coupled heroism and madness. It was a short step to the kind of move that pseudo-Aristotle makes in connecting melancholy with possessed heroes like Hercules and Ajax. What is more,

> the same is true of most of those who have handled poetry. For many such men have suffered from diseases which arise from this mixture in the body, and in others their nature evidently inclines to troubles of this sort. In any case they are all, as has been said, naturally of this character.[12]

Poets are moody, depressive types prone to introspection and inclined to melancholy because of the excess, or at least predominance of, black bile in their humoral constitution. This influential pronouncement combines that Platonic spark of divine poetic inspiration with the physiological basis of the melancholy humour to produce a figure who is both manically productive and yet depressive. He is not to be equated with the modern conception of manic depression. Nevertheless, the influence of *Problemata* 30 has reverberated throughout Western culture.

After Aristotle and his school came a number of writers on melancholia, including the Roman Celsus (*c.*30), Soranus of Ephesus (c.100), Rufus of Ephesus (c.100), Aretaeus of Cappadocia (c.150) and Galen (c.131–201). Galen we have examined already in relation to his refinement of humoral theory, while the others lent their own inflections to the broad classical discourses on melancholy as the product of black bile. Celsus, for example, apart from recommending the familiar treatments that might be supposed to remove excessive quantities of black bile—bloodletting, purging through hellebore, and exercise—advocated cures of a more psychological nature:

> Causes of fright [should be] excluded, good hope rather put forward; entertainment sought by storytelling, and by games, especially by those with which the patient was wont to be attracted when sane; work of his, if there is any, should be praised, and set out before his eyes; his depression should be gently reproved as being without cause; he should have it pointed out to him now and again how in the very things which trouble him there may be cause of rejoicing rather than of solitude.[13]

Some of these treatments seem modern, a variation of 'positive thinking' and distraction strategies. We should also note the recurrence of the idea that depression is excessive sadness, 'without cause', although one doubts that being 'gently reproved' for it would make much difference.

Rufus of Ephesus's work is notable because it had a powerful impact on ideas about melancholy, via his influence on Galen, but also because it influenced later Arab physicians whose own views fed back into Southern Europe and thence to medieval medicine and culture across the whole of Europe in the medieval period. Writing in the early ninth century, Ishaq ibn Imran followed Rufus, and in turn Constantinus Africanus (d. 1087) reflected Imran's work in *De Melancholia*, a text he translated from Arabic into Latin and which was widely read through the Middle Ages and the Renaissance. Rufus went with other thinkers in stressing the importance of black bile in the production of fear and sadness, along with a host of other symptoms that would not fit into a straightforward modern definition of depression. He stated that some melancholics might have a single obsessive and delusional notion even while they were perfectly sane: one man believed he had no head, another that his skin had become as dry as paper, and one thought he was a pot. He also accepted the fact that some melancholy people could have the gift of prophecy—a suitable corrective of the impression sometimes given that 'rational' Greek medicine was a straightforward precursor of modern science. As we see in the creative literature of the period, religious causes of mental illness were often paramount in classical culture.

Rufus also enhanced the Aristotelian idea that melancholy was the disease of the scholar: 'much thinking and sadness cause[d] melancholia'.[14] This is a different inflection from the *Problemata* 30's thesis, however, as it was *mental* activity that would bring on melancholy, not black bile in this particular instance. A desire for solitude and paranoia was often an attribute of the melancholic; no doubt this quality fed into the life of the studious. As with pseudo-Aristotle, Rufus also distinguished between those who are melancholic by natural tendency and those who become melancholic due to lifestyle factors such as eating too much 'heavy' meat (cows and goats) and 'thick and dark wines'.[15] Rufus pointed out that melancholia occurred in both genders as well as infrequently in the young and very commonly in the old. Men were more usually afflicted in general, although women suffered more severely when they did fall under its influence. Whatever the gender, it was important to nip the disease in the bud, because the longer it went on without intervention, the worse the melancholia could become.

Rufus and the other physicians mentioned here helped shape Galen's humoral ideas about melancholy and the Hippocratic tradition in general that were to persist, with the aid of systematisers of the Arab world (Rhazes, Avicenna, and Byzantine medicine) for centuries to come. The medical theory of melancholy remained remarkably consistent from classical antiquity to the later Renaissance.

Christianity and acedia

The advent of Christianity brought with it a profound shift in medical thought, even though Galenism persisted in a variety of forms. The isolation of Egyptian desert monks in the fourth century gave rise to a set of symptoms that included a nostalgia for their previous lives and a hatred for the present monastic one, low mood, ennui, and general misery. The term *acedia* (later *accidia* or *accedia* or, in English, *accidie, accydyei, acedy*) was taken from the Latin to indicate this condition. Such a depressive illness was framed by the specifically Christian context of the struggle against the worldly temptations of the devil and the sins of the flesh, and the related pressure to attain the unattainable: a spotless soul. One can see that the sin of despair might not be far away from such self-imposed discipline, and indeed that sin was mentioned in conjunction with acedia. The concept of acedia made its way into the West, not least because similar pressures faced all God-fearing Christians in a broader sense, whether they were confined in a monastery or convent or not. Saint John Cassian (*c.*360–435), known as one of the 'Desert Fathers' who inhabited the Scetes Desert in Egypt in the third century, wrote 'Of the Spirit of Accidie', which describes the 'combat' with the 'vice', where it is compared with 'the mid-day demon' spoken of in the ninetieth Psalm.[16] The noon-day demon might be the acedia or the sinful thought inflicted on the sufferer. In later medieval literature (like Chaucer, Gower, and Langland) acedia becomes a form of sloth and, as such, is a sin of idleness. As Chaucer's Parson puts it in the *Canterbury Tales*, 'Accidie' brings 'sloth' and 'hevynesse' or 'sadness' (which in Middle English meant both physical and mental depression), and is like being in the 'pain of hell'.[17] The crucial aspect of acedia in relation to depression is that we can only understand it via the religious framework of medieval culture.

Melancholia persisted alongside acedia according to the Galenic definition of the melancholic, clearly with some over-lap in the often severe dejection and even suicide brought about by both conditions, but the melancholic tended to suffer from delusions brought on by the malfunctioning of the black bile and its consequent vapours clouding the brain. These delusions provided a dividing line between the two illnesses, while it was also the case that melancholia, having a physical basis, was less stigmatised than the moral disorder of acedia, and might be a preferable diagnosis for the sufferer. Later there were attempts to define acedia within the medical model in relation to both the humours of phlegm and black bile. In either case, however, cure for disease in the Middle Ages involved a two-pronged attack on the soul and body, with a combination of prayer, purges, and drugs thought desirable to cure melancholia as well as acedia.

Acedia faded from the constellation of melancholy's cog-nates with the arrival of Renaissance classicism and reli-gious Protestantism, both factors that tended to weaken the Catholic Church's influence on European consciousness and the emotional states of the northern inhabitants in particular. The dissolution of the monasteries in England under Henry VIII, for example, attacked the origins of acedia in that lifestyle, but already the aspect of acedia denoting sadness was beginning to be absorbed into melancholia, along with a tendency for Latin terms like *tristitia* (the cardinal sin of sadness) to be rendered as 'melancholy', or 'melancholia' in the English version. In the increasingly differentiated religious framework of the Renaissance, acedia came to seem more and more like a relic from the Middle Ages, and the stage was set for the rise of Melancholy as a powerful cultural force.

II

~~~

GENIUS AND DESPAIR

When I go musing all alone
Thinking of divers things fore-known.
When I build castles in the air,
Void of sorrow and void of fear,
Pleasing myself with phantasms sweet,
Methinks the time runs very fleet.
…
All my joys to this are folly,
Naught so sweet as melancholy.
…
'Tis my sole plague to be alone,
I am a beast, a monster grown,
I will no light nor company,
I find it now my misery.
The scene is turn'd, my joys are gone,
Fear, discontent, and sorrows come.

'Abstract of Melancholy', Robert Burton,
The Anatomy of Melancholy, i, p. lxix.

Robert Burton (1557–1640), a solitary scholar at Oxford University, wrote the most famous and encyclopaedic treatment of melancholy in the Renaissance. He was, perhaps inevitably, a melancholic himself—the profession of the academic, with its sedentary lifestyle and overemphasis on abstract and intense speculation, leading to a humoral imbalance that could result in melancholy. Burton's poetical 'Abstract' repeats a paradox that we saw *in utero* in the classical period: melancholy can be a condition to be enjoyed, a 'sweet' disease in which one can luxuriate, a meditative mood more than a problem. This initial enjoyment of solitary speculation does not continue, however, and Burton soon finds that melancholy is Janus-faced, in which pleasant 'musing' transforms suddenly into misery. The melancholic becomes a 'monster' who avoids company, is drawn towards darkness, and with whom joy turns to the typical symptoms of classical melancholy, and those of modern depression: fear and sorrow.

The Renaissance saw the rise of the first form of melancholy in a flourishing of the myth of melancholic genius that has persisted up to the present day. Our emphasis in this book is on the second form, the serious disease that can derail or even destroy lives, yet there is a tension or at least a relationship between the two forms, a shadowy borderland between pleasure and pain, health and illness, creativity and paralysis, hope and despair. To understand where we stand with depression in the twenty-first century, and recent attempts to recuperate depression as melancholy, it is necessary to understand and acknowledge the presence of the more positive discourse of depression, however paradoxical it might be. This second tradition is often known as the Aristotelian mode, derived as it was from Aristotle's pronouncements on the genius of the melancholic, while the first is more broadly dubbed Galenic, because it takes Galen's more sceptical and pragmatic approach to melancholy as an illness to be suffered rather than enjoyed. In practice, the two modes manifested themselves in a more complicated way than this broad division suggests:

the celebration of melancholic genius could be blended with a lament for the heavy suffering the disease caused, because Galenic ideas about the misery of melancholy were mixed with Aristotelian notions of creativity—melancholics could be 'gifted and sick'.[1]

Renaissance definitions of melancholy

The early modern period witnessed a shift in the perception of melancholy from the Galenic humoral model to the New Science, in which alchemy changed to chemistry and in which the humours were replaced by nerves, spirits, and fibres. The balancing act of the humours to keep the body healthy was replaced, gradually and unevenly, even into the nineteenth century, by different conceptions of a balanced bodily econ-omy based on more mechanistic and chemical ideas of flows that might be blocked, or chemical imbalances that should be corrected. Like the melancholy of the classical period, early modern melancholy in the Galenic stage was considered to be generated by the overheating of black bile and the consequent fumes or vapours ascending to the brain, as Burton, citing Galen, commented: 'the mind it selfe, by those darke obscure, grosse fumes, ascending from black humours, is in continuall darkness, fear and sorrow, divers terrible monstrous fictions in a thousand shapes and apparitions occurre, with violent passions, by which the braine and phantasie are troubled and eclipsed'.[2] The melancholy poet and priest John Donne com-plained: 'But what have I done, either to *breed*, or to *breathe* these *vapors?* They tell me it is my *Melancholy*.'[3]

This major symptomalogical difference between melan-choly and depression persisted from Galen to Burton, but its effects in disordering the imagination were interpreted as being more or less severe, depending on whether one agreed with Aristotle's mythology of melancholic genius or Galen's less optimistic view of melancholy as a serious physical and mental illness. That said, Lawrence Babb has pointed out that both traditions were 'hopelessly entangled in Renaissance literature', as the terminally miserable but scholarly and refined genius of Shakespeare's Hamlet illustrates.[4] The experience of melancholy in the world beyond that of representations on the stage or in poetry could be another matter again, but one not necessarily separate from the kinds of models of melancholy behaviour being proffered in these literary sources. It is one of the arguments of this book that melancholy and depression are shaped by their cultural contexts to a greater or lesser extent, and that those cultural contexts include works of creative literature and art, as well as broad discourses of religion, class, and gender, all of which were prominent factors in the making of melancholy in the early modern period.

Although we are tracing an idea of modern depression throughout this book, and such an object seems to exist in all the periods we are examining, we must also emphasise the at times alarmingly vast cultural discrepancies between our contemporary view of depression—and the culture that both surrounds and constructs it—and that of the Renaissance. The shift from religious to secular, from humoral to scientific, from prayer to drug, had not yet occurred in this period, and the understanding and treatment of melancholy shuttled between the religious, magical, mystical, alchemical, demonological, and the more naturalistic explanations proffered by Galenism and the move towards various forms of humanism. This period saw no problem in treating melancholy with both religious and physical cures. Despite the profound strangeness of these cultures of melancholy depression, we can still see human suffering at the core of these narratives, but we should also not ignore the foreignness of expression at the same time.

Hannah Allen (*c.*1638-?), for example, was a Presbyterian writing about her trials of the spirit toward the end of our period in the 1660s, and described the way she viewed her own illness: 'As my Melancholy came by degrees, so it wore off by degrees, and as my dark Melancholy bodily distempers abated, so did my spiritual Maladies also, and God convinced me by degrees; that all this was from Satan, his delusions and temptations, working in those dark and black humours, and not from my self.'[5] Allen's interpretation of her depression was almost entirely dictated by a religious framework: it had room for physical understanding and physical treatment, but the dominant mode of her *Narrative* placed it firmly in the genre of spiritual autobiography. For her, and others, depression was part of a spiritual journey towards salvation, redemption, the gaining of God's grace, rather than a secular and humoral explanation. It is this alien nature of cross-cultural depression that helps to shed light on the particularities of depression in our own time.

Galenism and gloom

We turn to Galen first as the (at least partial) counterpoint to the Aristotelian and later Ficinian idea of melancholic genius. Although it might seem that there was an unchanging Galenic idea about melancholy since the time of the Ancients to the birth of the New Science in the seventeenth century, there were several shifts and developments that meant new ideas about the nature of melancholy and depression arose

in the early modern period. Galen and his '-ism' were useful and flexible reference points for the early modern world:
the 1525 edition of his works led to a renewed interest in his
thought across Western Europe and, because Galen and his
followers were not always consistent in their observations and
theories, they made good fodder as justification for a number
of contemporary religious and political ideologies. Galen had
also embedded the humours more profoundly in the actual
matter of the body than had Hippocrates. Galenism seemed
to have the answers to everything, from the activities of the
body to those of the soul, if one were prepared to ignore its
contradictions.

As we saw in the last chapter, Galen and his tribe had ascribed
the production of melancholy to an imbalance in the four
humours, specifically the black bile. 'Natural melancholy' was
present in the body and could be healthy, but if that natural melancholy were to become burnt or 'adust', or if the natural melancholy were to become excessively present for some reason, then
depression (causeless fear and sorrow) and a variety of other
symptoms both physical and psychological might result, such
as hallucinations, languor, sleeplessness, an attraction to solitude, and even rejection of society in general. The burnt 'choler'
had a complication: it could be very hot or very cold, and each
state affected men in a different way: when hot, it made men
'madde', furious, excessively merry, or lustful; when cold, 'dulle'
and stupid. So when melancholy arose from an adust humour,
it had a hot stage followed by a cooling which then went into
a cold one. To add to the complexity of the disease of melancholy, different humours could be burnt, leading to different
melancholy effects. Burnt natural melancholy would cause fear
and sadness; choler would cause furious raving; blood would

46

lead to undue levity, and symptoms could mix if more than one humour were burnt together.

One could be temperamentally of a melancholic disposition, which was not a disease, or one could be tipped over into full-blown illness. Black bile retarded and depressed the vital spirit that animated the body and the animal spirits that connected the mind and senses. Black bile clogged the animal spirits in the blood 'with the fogge of that slime, and fennie substance, and shut up the hart as it were in a dungeon of obscurity, causeth manie fearfull fancies, by abusing the braine with uglie illusions, and locketh up the gates of the hart…whereby we are in heaviness, sit comfortless, feare distrust, doubt dispaire, and lament, when no cause requireth it'.[6]

The melancholy man or woman of the Galenic tradition is not likely to be a success in society because melancholy 'taketh away from a man his sharpness of witte and understandinge, his assured hope and confidence, and all his manlye strength and courage, so that he hardly eyther attempteth or atchieveth any matter of excellency and worthyness: for such be doltish, slow and lumpishe'.[7] At this point the adust melancholy has cooled down: when hot it had other effects more akin to mania. This reduction of man to idiocy is not helped by the associated aversion to society—dark in humour, dark in personality goes the analogy: melancholics 'shunne the light, because that their spirits and humours are altogether contrary to the light', and wear their hats over their eyes.[8] They are drawn to 'desarts and solitary places, where they confine themselves, and consume themselves with the discontent and *Hatred* they beare to mankind'.[9] As a consequence of the mental agonies they undergo, melancholics 'weary of their lives, and feral thoughts to offer violence to their own persons come into their minds' and 'many kil and destroy themselves'.[10] Suicide was the end of the road for the melancholic persecuted by the black bile.

Melancholics had a matching image to go with their physical condition, or rather an image that was directly constructed by their physical condition. They were hollow-eyed, skinny, looked depressed, downcast, shy, and slow and silent. They were also hairy and had a complexion to suit the bile: 'of colour black and swart'.[11] This was an image that could quickly tip over into stereotyping and satire, but there was nevertheless a firm basis for it in the medical literature. The 'Goths' and 'Emos' of contemporary youth culture have inherited this melancholic ambience, although filtered through later melancholic traditions such as gothic Romanticism.

Galenic representations of depressive melancholy in literature emphasised the role of heat and moisture in the body. If a person were suffering from prolonged or extreme sorrow for whatever reason (like the death of a family member or lover), then the body would be drained of its natural heat and moisture which are contained in the blood's vital spirits, thus affecting the vital spirits themselves and causing the blood to become cold and dry melancholy. This pathological melancholy would waste and eventually kill the affected person: 'Dry sorrow drinks our blood' says Romeo to Juliet when they separate for the final time (III, v, 59). When we turn to notions of melancholic genius, we see that a balanced heat is necessary for the image of melancholy to escape its Galenic gloom.

Women, in the Galenic model, suffered in the same way as men for the most part: humours were common to both sexes after all. The womb was a complicating factor, as it could lead to humoral imbalance and the burning of black bile if women were not sexually satiated and the menstrual blood duly evacuated. Earlier in the period, women were thought to be especially lustful and in need of male semen to restore the correct balance of heat and moisture. In representation at least, the broader patriarchal stereotypes of women as disruptive and passionate (as opposed to rational men) dovetailed with the Galenic conception of the dangers of the female reproductive system to bring disorder and melancholy. Ophelia's frustrated love for Hamlet can be interpreted in this way to some extent. Representation was one thing, however, and the actual treatment of women was another.[12] Katharine Hodgkin has shown that gender difference or indeed sexuality was not necessarily invoked in some of the cases she examines from the seventeenth century, and that religion might 'trump' medicine and, even when medical language was used, it could be 'asexual'. Galenism, however, was not the only theory of melancholy, and women had a further battle to fight in the Aristotelian mode.

Renaissance genius: Aristotle to Ficino

'Melancholy Men Or All Others Are Most Witty, [And Their Melancholy] Causeth Many Times Divine Ravishment, And A Kind Of *Enthusiasmus*…Which Stirreth Them Up To Be Excellent Philosophers, Poets, Prophets, Etc.'

(Burton, *Anatomy*, i, 461)

We saw that Aristotle, or at least pseudo-Aristotle, put forward the contentious but highly influential concept of the melancholic (male and upper class) genius in the classical period. In the Renaissance this rather paradoxical idea of disease as a vehicle for intellectual prowess and creativity was seized upon by the melancholic scholar Marsilio Ficino (1433–99) in Italy and thence disseminated across Europe, culminating in the dramatic

49

Fig. 4 'Melencola I' after Albrecht Dürer. This 1514 engraving captures the personal misery of melancholia while representing the social and philosophical consequences. Knowledge of all kinds grinds to a halt, and life itself seems futile. Dürer's picture was extraordinarily influential for future images of melancholy. (*Wellcome Library, London*)

character of Shakespeare's Hamlet. Ficino's Neoplatonic astrology identified the star sign of Saturn as melancholic (as in Dürer's *Melancholia*) and decreed all men of genius and learning to be melancholic and under Saturn's influence in one way or another (see Fig. 4).[13]

Ficino did not arrive at the idea of melancholy as a positive condition without a large body of ideas from different discourses informing his own decisive formulation. A number of factors led to this apparently unlikely embrace in Ficino's time of the concept of 'genial melancholy', including the related discourses of love (earthly and Platonic), religion in its many forms, alchemy, humanism, the role of the scholar, class, and, of course, gender. These elements of melancholy in this period are important, sometimes complicated, and bear further examination.

Love and melancholy

Love was an important aspect of melancholy in this period, more so than in early ones. It is a major category in Burton's *Anatomy of Melancholy*. Love melancholy, the desire of the lover for his or her beloved, had been a condition commonly recognised from classical times: violent passions could cause the black bile to burn more than it ought, with the resulting imbalance in the humours. Desire would increase the heat of the body, humours would burn and melancholy would ensue. The idea of a noble 'heroic love', in which the melancholy lover becomes an authentic representative of a higher form of love, emanated from 'a mere linguistic coincidence'. The Greek word *eros* meant love, and 'merely by aspirating the *e* in *eros*, the writer could transform the word into *heros,* the equivalent of the Italian word 'eroe' and English word 'hero'.[14] Because the black bile was kindled by the passion of love, the melancholy lover could be compared to the similar kindling of the melancholic ancient heroes like those mentioned in Aristotle's *Problemata*, 30. This was a sleight of hand, but nevertheless a convincing one to many in Ficino's time, where the influence of Plato had crucially introduced the idea of a progression from earthly love to one that is more divine and pure. It was a short step from the idea of heroic love to the concept of melancholics being attracted to divine wisdom as manifested in various forms of artistic beauty, be they poetry, painting or philosophy, or religion in general.

Medieval representations of love melancholy saw it as what we might now consider to be a manic-depressive process, with the process of falling in love being a burning of the black bile in a form of ecstasy, followed by depression and inertia when the object of that love cannot be grasped. It would be a mistake to consider this as in any way related to modern manic depression as the mechanisms and logic of the process of love melancholy are entirely different from the concept of manic depression as it arose in the nineteenth century. Michele Savonarola (1452–98), a Professor of Medicine, discussed the state of 'melancholic solitude' in which those 'being so disposed from a disordered love, are in a state of continuous thought, memory, and imagination'.[15] These sufferers are usually the more refined sort, and the passion 'is called *haereos*, because it more frequently befalls heroic or noble men'.

Melancholy lovers were attracted to the landscape of melancholy which was to become so familiar in later representations: one Italian physician points out that the heroic lover 'flees the conversation of men and seeks out lonely places among the sepulchres of the dead where, feeding up his solitary melancholy, he might perpetually meditate upon the beauty of certain

Fig. 5 A depressed scholar. Scholars were thought to be particularly prone to
melancholy due to their solitary, sedentary lifestyle and over-intense thought,
as this etching by J.D. Nessenthaler illustrates, *c.*1750. (*Wellcome Library, London*)

forms'.[16] Even before Ficino completed his Platonisation of the idea of melancholy, the possibility of a shift from love melancholy to religious melancholy is clear in passages like this. We have already seen that the *acedia* of monks was a recognised phenomenon in the Middle Ages, an indication that the quest for divine love might involve suffering for the soul.

Related to this form of love was the love of learning: an heroic scholar questing after knowledge, which might in itself be religious, could fall victim to melancholy (see Fig. 5).[17] Aristotle's list of melancholic heroes encompassed not just the likes of Hercules and Lysander, but also philosophers like Empedocles, Socrates, and Plato. Scholars too could overreach, wrecking their health seeking wisdom. One reason was that the medical writers associated the life of a scholar with melancholy, which might be seen as both a cause of depression and yet as an effect.

Religion and melancholy: the 'devil's bath'?

Melancholy posed a problem for the religious: was melancholy caused by demons and devils or did it cause them to appear as hallucinations? It is here that the alien nature of medieval and Renaissance culture becomes most apparent. Those suffering from melancholy were commonly said to have visions of various types and, in an intensely religious environment, it was no surprise that many of them were associated with holy (or unholy) themes, and continued to be so at least until the end of the eighteenth century. In the mid-sixteenth century Francis Spira, a wealthy Italian melancholic, suffered an attack of religious melancholy when he converted back to Catholicism, having originally converted to Protestantism: he stated 'that he saw the divels come flocking into his Chamber, and about his bed, terrifying him with strange noyses; that there were not fancies, but that hee saw them as really as the standers by'.[18] Spira regarded himself as being punished by God for his apostasy, but there were many explanations offered by different parties, depending on their politics, religion, and class. One of the tasks of demonologists was to distinguish the symptoms of a 'natural' melancholy from those produced by demonic interference. An odd by-product of this attitude to melancholy was that it left the way open for its opposite: rather than demons producing melancholy, God might see fit to use melancholy as a vehicle for divine genius, a mark of approval rather than the kind of condemnation reserved for the malign and devilish genius of witches and sorcerers.

Mysticism provided a further positive role for melancholy: melancholy could be identified with the 'dark night of the soul' through which the soul had to go in order to achieve purgation in its journey towards divine forgiveness and knowledge. Even as mystics tried to rid themselves of melancholy through spiritual exercises, they might also regard a certain kind of depression as inevitable or even desirable: to lack depression might mean that one was not doing enough to examine one's soul. St Caterina of Genoa (1447–1510) typifies the earlier attitude to melancholy, in which it represents the 'devil's bath', an obstacle to her salvation.[19] She was 'so stifled in a continuous state of melancholy [*malinconia*] that she did not know what to do with herself'.[20] She called to God to rid her of devilish melancholy, whereas later Ficino found a role for melancholy as a catalyst for the soul's ascent towards divine love. It had been noted by others, however, that Christ suffered from melancholy, and that the concept and image of the sorrowful Christ was in conflict with the Stoic attitude that one must rise above the passions and be indifferent to them. St Antonino praised the kind of sorrow emanating from 'a desire for the heavenly fatherland and from the love of God'.[21] Ficino would take this even further in his statement that melancholy could actually become a springboard for theological and spiritual speculation.

Humanism and melancholy

In some ways melancholy was antithetical to the social existence and civic responsibility being encouraged by the new discourse of humanism: melancholy encouraged solitude, misanthropy, and, even worse, an attraction to places that symbolised a rejection of mankind, such as cemeteries. Already one can see the connection with the Gothic of the late eighteenth century, in which the graveyard school of poetry would feed on the idea of melancholy.

The rise of city-based commerce in the early modern period and the move away from agrarian economics led to a new emphasis on an ideology of the city and the citizen based on the political theories of the Ancients—an emphasis that rejected contemplative solitude and embraced the civic responsibilities of the individual. Civic humanism, as we now know it, seemed inimical to a celebration of melancholy. Fortunately for theories of melancholic genius, there were fault lines in such an ideology of healthy sociability, not least in the person of Francesco Petrarca, or Petrarch, the famous poet (1304–75). He lauded solitude as necessary to his studies and lived a monkish lifestyle which resulted in mental problems that he compared to '*accidia*', the depressive languor associated with a monastic existence. Humanists understood that the heroic labours of the scholar might involve, perhaps even necessarily, a Promethean or Herculean melancholy (for these were their classical heroic models), suffering in order to achieve the goals of a civil society. A tension remained, however, between the melancholy of such heroes who were ostensibly bringing good to their fellow men, much as Prometheus suffered to bring fire to man, but they simultaneously showed that they were capable of going beyond the limits of the normal, the individual, and the state in their reaching for the infinite, a phenomenon related to heroic love.

Ficino and the glories of melancholy

Marsilio Ficino brought together the strands of thinking leading into the Renaissance valorisation of melancholy that we have mentioned, and he himself was a melancholic, which might explain his vested interest in painting such a rosy picture of melancholy, at least in its moderate form. In his *De Studiosorum Sanitate Tuenda*, the first book of *De Vita Libri Tres* (1482–9) he elaborated on the Aristotelian idea of the fine line between just enough black bile for creative genius and too much.[22]

Ficino stressed the need for scholars to control and foster their melancholy, and wrote poetically about the wonders of natural melancholy. According to Ficino, kindled melancholy—a gold and purple colour—radiates the colours like a rainbow, and generates spirits perfectly suited to mental activity. The connection of melancholy to the earth and Saturn meant that it both penetrates into the deepest secrets of the natural world and also aspires to the highest truth. Melancholics could even be conduits of divine prophecy, forming 'ideas never before conceived' and predicting 'events yet to come'.[23] As Du Laurens, one of the many Renaissance thinkers to echo Ficino's views on melancholy, put it: heated melancholy combined with blood 'causeth as it were, a kinde of divine ravishment, commonly called *Enthousiasma*, which stirreth men up to plaie the Philosophers, Poets, and also to prophesie: in such maner, as it may seeme to containe in it some divine parts'.[24] Poetry could emanate from God, said the Renaissance poets. It was a form of heroic love, a divine frenzy as Plato had said. Boccaccio saw the poet as a divinely inspired mystic who delivers truth to man through his poetic forms and symbols. From here Ficino made the connection with Aristotelian melancholy, and the image of the melancholy poet, so popular in later centuries, became cemented in the popular imagination.

Unlike Galen's rather closed and introverted perspective on the body, Ficino's emphasis on the Platonic and metaphysical world saw the body as being open to the influence of the stars, thus connecting the older concepts of melancholy genius with astrology, a science newly invigorated in this period by the increasing sophistication of the technology of telescopes. This was indeed a liberating move on Ficino's part, but why was Saturn the sign of the melancholic? At first glance, this planet did not seem a likely candidate to preside over genius. Astrology was an ancient science, but it took the Arabic thinkers of the ninth century to match humours to particular stars: Jupiter was sanguine; Mars was choleric; and Venus phlegmatic. Saturn was thought to be cold and dry, like melancholy, and its slow revolutions seemed to suit the melancholic temperament. It was also thought to be dark, which suited the black bile. The mythological origins of Saturn added a mix of sometimes contradictory qualities: Saturn was also Kronos, king of the Golden Age, but subsequently deposed by his son (and castrated in the process) and condemned to live in a gloomy subterranean prison. This exile also connected him with travel. There was also confusion with Chronos, god of time and master of birth and death. Through these origins Saturn was thought to command melancholy in all its aspects, including related physical elements like the spleen. He became a symbol of authority, good and bad, and in his worst representations he seemed allied to self-harm, drowning, and imprisonment.

For Ficino, Saturn became the planet of the most refined speculation and the ruling sign of scholars: the Platonic concept of heroic melancholy enabled him, together with Kronos's kingly status, to elevate Saturn into the seemingly paradoxical realm of unearthly and abstract spiritual contemplation. Saturn was an enabler of the greatest minds, but the melancholic had to beware of Saturn's great powers, which could be turned against those under its influence. Ficino devoted a great deal of effort to recommend ways to control Saturn's potentially deleterious effects on those born under its sign: not only diet, but also talismans that might counter Saturn with complementary qualities from other planets.

Alchemy is another occult and mystical discipline utterly foreign to us now, yet was influential in interpreting and constructing depression. The influence of Arabic medicine and philosophy was crucial: *al-kimiya* means 'alchemy' in Arabic, and Arab physicians were expected to have a working knowledge of the art, and its mission to transform elements into new substances, and finally base metal into gold. The quest for the essences of substances as part of this transformation was thought by some to be a sacred activity, as it required the revelation of hidden or occult truths about the world. Alchemy became tangled with Platonic ideas about ideal forms and also with astrological ideas about one's humoral temperament as being partly dictated by place and date of birth. Muslim culture helped perpetuate and integrate the theory of fundamental correspondences between the microcosm of earthly reality and the macrocosm of the stars and higher spiritual truths. Melancholy was identified with an initial stage of transformation called '*nigredo*' or blackness (of the black bile metaphorically), dominated by Saturn and analogous to temporary death before resurrection. Once this alchemical doctrine became fully connected with the astrological implications of Saturn as the dominant aspect for human genius, the Florentine revolution of Ficino would be underway.

Galenic versus Ficinian melancholy

The relationship between these two traditions—the misery of Galenic melancholy and the inspired genius of Aristotle and Ficino—is complicated, to say the least. The two conceptions coexisted uneasily in Renaissance culture, although in the medical literature Galenism had the upper hand. It has been argued that there was a shift later in the Renaissance from the more ethereal Neo-Platonism of Ficino and the poet Edmund Spenser to a form of Galenic humoral materialism (which was anti-Platonic). Galenism moved 'the locus point of melancholy' away from either love melancholy, in which the loved one is absent, or other situations involving loss, to within their own bodies, 'where it bubbled and burned in the spleen'.[25] The seventeenth century saw the Aristotelian glamorisation of melancholy tempered by a Galenic awareness of the forces of the black bile working within the self, a situation that encouraged intense introspection and awareness of the bodily boundaries of the self, and a certain alienation from that physical self. There was a broad distinction between a Galenism that 'renders the body as a more or less closed system: always already sick, or—if healthy—only temporarily so', and Neoplatonic approaches that 'emphasised the body's openness to cosmic ethers'.[26]

John Milton's pendant poems 'L'Allegro' and 'Il Penseroso' take both traditions and contrast them: the former seeks to reject the Galenic depression that crushes creativity:

> Hence loathed Melancholy
> Of *Cerberus*, and blackest midnight born,
> In *Stygian* Cave forlorn
> 'Mongst horrid shapes, and shrieks, and sights unholy,
> Find out some uncouth cell,
> Where brooding darknes spreads his jealous wings,
> And the night-Raven sings.[27]

The latter hails the life-affirming vision of Aristotle and Ficino: this Goddess is a 'pensive Nun, devout and pure,/Sober, stedfast, and demure' (l. 31–2) who is Saturn's daughter and

> Whose Saintly visage is too bright
> To hit the Sense of human sight;
> And therefore to our weaker view,
> Ore laid with black staid Wisdoms hue (l. 13–16).[28]

Milton ends the poem by choosing to live with this version of melancholy: an acceptable, indeed desirable celebration of the poetic, scholarly and pious life. For Milton, melancholy might be containable, even inspirational, even as he recognised the damage it could do in its more severe form.

Towards social fashionability

The glamour invested in melancholy by the Aristotelian tradition meant that men and women were willing to announce themselves as having at least a touch of the melancholic disposition or 'habit'. Although the vogue for melancholy began in Italy, it quickly spread through the network of European intellectuals and aristocrats to the north, and became firmly entrenched in Britain, to the extent that characters like Hamlet manifested themselves in literary representations, and then so firmly stereotyped that satirical portraits of moody young men started to appear shortly afterwards. Hamlet himself is subject to a certain amount of black satire. The melancholy 'malcontent', usually a disaffected young man frustrated in his progress up the slippery ladder of ambition, became a recognised social type by about 1580. The individual in question would have at least a self-image of talent, if not genius, but also a perception that the world, for reasons of its own, was refusing to recognise that Saturnine blessing of great ability.

Literary representations mocked social ones, albeit in an exaggerated manner: in John Webster's revenge tragedy *The Duchess of Malfi* Antonio accuses the malcontent Bosola of affecting an 'out of fashion mellancholly' (II, i. 95). This was written and performed around 1613–14, so we might assume that trends had moved on. As we see throughout this book, however, melancholy and depression have had a tendency to be linked to some forms of social cachet with an apparently surprising tenacity: a poem in the 1630s represents the writer John Ford as standing alone with 'folded arms, and melancholy hat', the traditional pose of the melancholy man (see Fig. 6).[29]

Melancholy was *de rigeur* for the fashionable, and the country bumpkin Stephen in Ben Jonson's *Every Man in his Humour* (1598) considers it to be an essential disease for his social advancement. Similarly, Lyly's *Midas* (1592) shows Motto the barber being satirised for saying that he is melancholy rather than 'doltish' because 'melancholy is the crest of Courtiers armes'

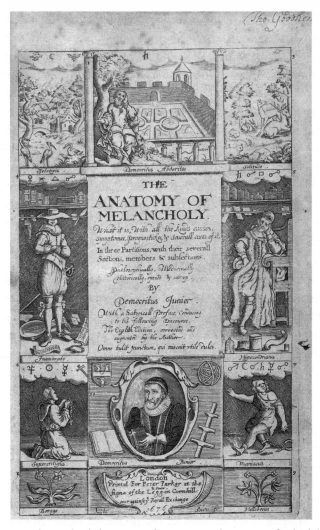

Fig. 6 A love melancholic. Burton's frontispiece to the *Anatomy of Melancholy* shows a male love melancholic or 'inamorato' posing on the left, along with other manifestations of the many varieties of Renaissance melancholy. (*Wellcome Library, London*)

(5.2.99–110). Melancholy in its social role needed props: 'have you a stoole there, to be melancholy upon?' (*Every Man in his Humour*, 3.1.100). Ironically, he becomes truly melancholic when he is mocked and swindled.

Although we need to be aware of the distinction between literary representation and social reality, it is appropriate that the role-playing of drama mirrors—and even to some extent constructs—the template for behaviour prompted by a condition that had become fashionable. No doubt all disease is in some way socially constructed, but the fashionability of melancholy in this (and other) periods is particularly illustrative of the way in which narratives of disease are built partly upon cultural representations in literature and the visual arts.

On the subject of social categories of disease being represented in literature, it is difficult to ignore the famous example of Jacques in Shakespeare's *As You Like It* (1599). He has clearly embraced a fashionable role that suits his sense of himself as heroically melancholic, even if he has a sense of irony about his—and the human—condition simultaneously:

> ROSALIND. They say you are a melancholy fellow.
> JACQUES. I am so; I do love it better than laughing.
> ROSALIND. Those that are in extremity of either are abominable fellows, and betray themselves to every modern censure worse than drunkards.
> JACQUES. Why, 'tis good to be sad and say nothing.
> ROSALIND. Why then, 'tis good to be a post.
> JACQUES. I have neither the scholar's melancholy, which is emulation; nor the musician's, which is fantastical; nor the courtier's, which is proud; nor the soldier's, which is ambitious; nor the lawyer's, which is politic; nor the lady's, which is nice; nor the lover's, which is all these; but it is a melancholy of mine own, compounded of many simples, extracted

64

from many objects, and, indeed, the sundry contemplation of my travels; in which my often rumination wraps me in a most humorous sadness.[30]

Jacques delineates various categories of melancholy, many of which are described in the medical literature, but goes on to claim that his own 'humorous' melancholy is unique, which, in terms of humoral theory, is true. Today Andrew Solomon blames his 'serotonin' (or lack of) for his depression, but in the Renaissance everyone is different in their illness, even if one could blame the black bile in a similar manner to Solomon's serotonin.[31] Each person has a slightly different balance of humours, and indeed each person has different experiences in their lives, exemplified here in Jacques's 'travels' and his alchemical metaphor of melancholy 'compounded of many simples'. Although a variety of categories existed to define melancholy in this period, Jacques illustrates the strong idea of the individuality of disease—possibly something that has been lost with the later advance of biomedicine. The example of Hamlet, as we will see later, gives us a further and greater instance of this drive to complexity in the understanding and representation of the melancholic.

Female genius and suffering

It was not only men that felt an affinity with melancholy genius and the gain to be had from its fashionability, however. Until recently critics have regarded women as being excluded from the masculine tradition of inspired depression propagated by Aristotle and Ficino, but closer examination reveals a number of female writers and thinkers who saw themselves as melancholic by temperament and used this as a justification for their forays into the intellectual realm at a time when any female attempt at the world of the mind was regarded with suspicion and often downright hostility.

Naturally those at the upper end of the social scale had greater access to an education that could equip women to take advantage of the pose of the melancholic scholar, even if they were not necessarily prone to the condition in real life. Margaret Cavendish (1661–1717) was the Duchess of Newcastle, and a prolific writer in all manner of genres, including natural philosophy (nowadays called science). Cavendish was duly attacked for her presumption to encroach on male territory: 'mad, conceited and ridiculous' said Samuel Pepys. Cavendish assiduously presented herself as a melancholic figure, very bashful, and a lover of books and solitude. Her frontispiece (see Fig. 7) to *The Philosophical and Physical Opinions, written by her Excellency, the Lady Marchionesse of Newcastle* (1655) has a poetic caption that explains the melancholic pose struck in the image:

> Studious She is and all Alone
> Most visitants, when She has none,
> Her Library on which She looks
> It is her Head her Thoughts her Books.
> Scorninge dead Ashes without fire
> For her owne Flames doe her Inspire.[32]

No doubt female authors had genuine causes for depression, given the frequently hostile opposition to their activities, even if—as was also often the case—they had male and female figures in their lives who did encourage their scholarly activities. Pepys's reaction is a case in point, but here Cavendish invokes a certain kind of 'madness' that actually validates her as an intellectual of distinction, and draws on the iconography of the

Studious She is and all Alone
Most visitants, when She hag none,
Her Library on which She looks
It is her Head her Thoughts her Books.
Scorninge dead Ashes without fire
For her owne Flames dos her Inspire.

Fig. 7 Margaret Cavendish, Duchess of Newcastle, who exploited the fashionable image of the intellectual male melancholic for women in this frontispiece to her *Philosophical and Physical Opinions* (1655). (*By permission of the Master and Fellows of St John's College, Cambridge*)

melancholy scholar, sitting alone with the tools of her trade: desk, pen, ink, books. The picture is clearly a form of role-playing, and yet its goal is to indicate authenticity of genius. Not only did women use the Neoplatonic and Petrarchan traditions to show women as rational or 'to depict their lovesick passion as the celestial vision that inspires them to a heroic quest', but also that male melancholy could be depicted as destructive, feminising, and, moving back to Galenism, a result of the retention of semen.[33] We will see that female intellectuals continue to tap into the possibilities for self-expression opened up by melancholy genius in the following centuries.

Although the theory of the humours could, as with other philosophies, be turned in the service of patriarchy, the renewal of Galenism also left an opportunity for women to argue that they too possessed black bile, and that they too could be possessed by it for the forces of good and intellect rather than merely as deluded witches. It has been argued that the Aristotelian mode excluded women, but this ignores the importance of the humours and Galen in the construction of melancholy. Clearly many women were debarred from the educational opportunities that might enable their participation in scholarly melancholy, but the fact that women were thought of as cold and dry in a humoral sense meant that they could be considered as melancholic. Medieval wood-cuts represented melancholy as either a despairing scholar or a female spinster. In this case Cavendish was late enough to be present at the birth of the New Science, in which such discoveries as Harvey's circulation of the blood destroyed—not instantly—the notion of the four humours, but it took a long time for popular iconography to catch up with such developments, as indeed did many medical treatments. The humours took a surprisingly long time to shift as a theory and practice in medicine and wider culture.

Some say that the positive perception of female melancholy went backwards when a new form arose in the seventeenth century that was related to disorder of the female reproductive organs, but we have seen throughout the history of melancholy—and will see in the history of depression—that sexual difference has been continuously reformulated according to developing notions of the body and utilised against the idea of female intellectual equality.[34] Even for the compendious Burton melancholy is masculine, and he only discusses women in one chapter, added in 1628, not in the 1624 original: 'Symptoms of Nuns, Maids and Widows Melancholy, in body and mind, &c.' Women here are tied firmly to their (frustrated) wombs, the source of disruption that leads to a specific form of melancholy. Yet women battled to overcome such obstacles, and attacked the supposedly exclusive idea of Renaissance masculine genius with ingenuity.

Melancholy was a mixed blessing for women, however. As Katharine Hodgkin has shown, we need to consider the various discourses in play when analysing how women dealt with melancholy. The Oxfordshire gentlewoman Dionys Fitzherbert attempted to distinguish her own religious struggles from 'melancholy or I know not what turning of the brain' in her autobiography in manuscript finished by about 1610: *An Anatomie for the Poore in Spirit, Or, the Case of an Afflicted Conscience layed open by Example*.[35] It was important for her to show that her melancholy had a good cause, and she made strenuous efforts to aid the godly reader in distinguishing a melancholy mad person— which is what the religious sufferer looks like—from one of God's elect enduring the necessary strife between good and evil,

God and the Devil, in her soul. The difference could be found in the humility and self-reproach of the godly, while the ungodly mad turn their rantings against other people.

When it came to the practical treatment of women on the ground, even the medical language that might appear to fix women in a position of disadvantage regarding melancholy could be trumped by religion. Melancholy might require both religious and physical treatments in fact, and the reality was that both were used to cover all bases: nor were religious and physical treatments (not greatly changed since the time of the Ancients) thought to be mutually exclusive. In these cases the dominant discourse was not the inspired genius of Ficino as invoked by the likes of Margaret Cavendish, but treatments motivated by the humours of Galen and the particular religious group to which the sufferer of melancholy belonged.

Hamlet: tragedy and genius

Robert Burton may be the most famous actual melancholic of the Renaissance, but Hamlet is the template of the melancholic man of distinction for future ages, just as Ophelia becomes the archetype of female lovesickness, a blend of passionate sexual desire and virginal purity that ends in suicide. Much has been written about Hamlet and his relationship to both traditions of melancholy, the more grim Galenic inheritance and the genial melancholy of Ficino and Aristotle—Hamlet is a complex blend of the two. What is clear is that Hamlet benefits from the genius of Shakespeare (whether we deify Shakespeare himself or not) and his ability to portray human complexity in a dramatic character, and nowhere is Shakespeare more successful in this endeavour than in the character of Hamlet.

Of course, we know that Hamlet is not a real person: he is a simulation, but Shakespeare draws together various discourses on melancholy that have had relevance not only for his own time, but also for succeeding generations. This is not to say that what Shakespeare writes is 'timeless', more that what Shakespeare invokes in terms of melancholy has continuing import in the Western tradition and regarding the condition itself, and in how we might think of depression as it follows on from the passing of melancholy.

Hamlet is a melancholic scholar in the Ficinian mode at the start of the play: although he plays with some stock roles of the melancholic à la Jacques in *As You Like It* (disillusioned scholar, satirist, misogynist, political malcontent, melancholy lover, madman), his melancholy is authentic and partly predicated on the general, existential sense of a melancholy universe, a world view that suits the genre of tragedy. The redemption of religion is not a factor in tragedy either: religious melancholy does not exist here because God does not exist here. Or if he does, it is as a vengeful deity who spares neither innocent virgin nor moral scholar. Hamlet himself becomes the primary revenger, of course, as the first title of the play suggested.

The wordy, speculative, questioning nature of Hamlet (he dominates the play in terms of lines and monologues) interrogates the place of humanity in a corrupt and hypocritical environment: 'to be, or not to be' indeed. Ficinian optimism has no place in this genre, however, and a Galenic awareness of the black bile burning the body undercuts the lofty power of philosophy, language, and religion. Here is the influence of Saturn in his malign aspect: Hamlet's intellect is a superior gift, and his sense of morality reaches beyond the mere 'seeming' of social politeness that masks hypocrisy, but, as his meditation upon Yorick's skull emphasises, he cannot transcend the earth, dust, and death.

Conclusion

This chapter in depression's biography has shown us a maturing concept in a world both familiar and unfamiliar. Familiar in that men and women of all classes displayed symptoms of depression, sadness, and anxiety, and yet unfamiliar in the way they interpreted those symptoms, and the way in which other symptoms were wrapped in with them. In this long period, the entire cultural understanding of medicine, body and mind, is alien to our contemporary world view, even if we can discern similarities with modern symptoms. Hamlet's literary example is a case in point: he seems to be a depressed man, but that depression is informed by ideas from Aristotle, Plato, Ficino, Galen, astrology, humoralism, humanism, alchemy, and so on. Melancholy was complex in its motivations, physiology, and cure, and the period coped with it in its own manner, however successfully.

Next we leave the world of the humours, and arrive at the New Science, a break with the world of magic, occult learning, and superstition that would ultimately, if not rapidly, lead to the modern world of biomedicine that we inhabit today, and with it a very different understanding of depression.

III

———— ∞∞ ————

FROM SPLEEN
TO SENSIBILITY

I feel my verse decay, and my crampt numbers fail.
Thro' thy black jaundice I all objects see
As dark and terrible as thee.
My lines decry'd, and my employment thought
An useless folly or presumptuous fault.
                    Anne Finch (1661–1720), Countess of Winchilsea,
                              'A Pindaric Ode on the Spleen'[1]

lthough Anne Finch describes both the travails of being depressed and the related pressures of being a female poet, however aristocratic, the moody and melancholic Hamlets and suicidal Ophelias of the Renaissance largely gave way to new, more sympathetic versions of the melancholic man or woman in the age of Enlightenment. Melancholy gradually became an aspect of the disorders of the nerves and, in England, a defining characteristic of English civilisation and its accompanying lifestyle as well as its climate: the *English Malady* as it came to be known. The core symptoms of sadness and anxiety persisted, but were framed and narrated by a new set of medical understandings and social circumstances, including mechanical ideas about the composition and functioning of the body, and an apparently more secular society in which the doctor became more important than the priest as a means of combating the dark night of the soul that had previously constituted Renaissance and medieval melancholy. Burton's *Anatomy of Melancholy* seemed an oddity to many now (although not to Dr Johnson, who found its engagement with religion an issue in his own life).

As in the Renaissance and the classical periods, there were also at least two competing versions of melancholy: the serious depressive form and the lighter, more positive type that had come to be associated with genius and heroism. Again, these traditional discourses of depression took new forms and had new implications for the sufferers of the condition.

Medical definitions

The late seventeenth century brought with it new problems of definition and nomenclature that grew out of the new medical understandings of melancholy: 'Before the time of Queen Anne, *melancholy*, as the name for morbid depression, had been largely replaced by *hypochondria*, *spleen*, *hysteria*, *and vapours*, all four terms denoting the same disorder.'[2] Even hysteria had become less gendered and more merged with melancholy. How did this happen?

The answer lies partly in the rise of the New Science, as we know it now. With a general cultural drive to investigate nature directly, rather than uncritically accept the wisdom of the Ancients, came advances in medicine and science. Thomas Willis (1621–75) followed the usual definition of melancholy as 'a raving without fever or fury, joined with fear and sadness' and then divided it into two forms: one in which the sufferer is universally delirious, 'so that they judge truly of almost no subject', and one in which the delirium is partial, and judgement 'amiss' in one or two cases.[3] In his *Two Discourses Concerning the Soul of Brutes* (1672) he broke with humoral theory, clearly denying that 'melancholy doth arise from a *Melancholick* humour' (*Two Discourses*, 192).

Willis himself followed the iatrochemical school, in which alchemy was becoming chemistry, and considered that the basic chemical components of the human body as then understood would cause health or disease. Melancholy was a result of malfunction of the animal spirits, which would ordinarily be 'transparent, subtle, lucid': in melancholics they became 'obscure, thick, and dark, so that they represent the Image of things, as it were in a shadow, or covered with darkness' (*Two Discourses*, 188).

Animal spirits, themselves to be thought of as similar to 'some Chymical Liquors', were transmitted via the blood to different parts of the body, including the brain, which could then be corroded by the badly fermented and vinegarish nervous liquor of the animal spirits. The spleen as a source of melancholy came into play when its normal role as a beneficial processor of blood by fermentation malfunctioned, and sent bad blood into the system again. Although Willis still used the word 'humour', it now meant bodily fluids, with blood pre-eminent after William Harvey's discovery of the role of the heart in pumping blood around the body. The old four humours still existed, but became waste matter from the blood, far from their previously lofty status of the lords of health and disease.

Willis's iatrochemistry did not radically affect traditional methods of cure, and used purging and bleeding to remove the supposedly bad matter from the body. The rise of the spa treatment, so famously associated with the eighteenth century, had its Protestant origins in the link of the spa waters with various minerals, like iron, that could strengthen the nervous juice or fluid. Catholic spa waters claimed miraculous powers, something the Protestant English avoided as Popish superstition. Tonics containing steel might also be drunk, for the same purpose of fortifying the animal spirits.

Iatrochemistry flourished briefly at the end of the seventeenth century, a herald of the new, non-humoral thinking to come. By the start of the eighteenth century proper, the new theory of mechanical philosophy and medicine had its impact on the concept of melancholy. These mechanical principles included hydraulic concepts made more urgent by the work of Harvey on the circulation of the blood. Archibald Pitcairn (1652–1713), the most important Scottish physician of his day and professor at Leiden before that, saw the body as a system of pipes or canals, whose job it was to carry the blood and its nutrients to the relevant parts. Disease was bad circulation of one sort or another: a blockage in the pipes, causing corrupt matter to accumulate and corrode. For Pitcairn, melancholy was caused by a slowing of the blood, a thickening and accumulation of the blood in the brain beyond the normal: the animal spirits would not receive such vivid 'vibrations' from the brain because of this dampening effect of the heavy blood, and thus result in melancholy.

Friedrich Hoffmann (1660–1742), professor of medicine at Halle University and famed across Europe, also focused on the movement of fluids through the body, although he did have a partial role for iatrochemistry through fermentation. Hoffman's views were highly influential in the first part of the eighteenth century: for him, melancholy was caused by bad or overloaded quantities of blood in the brain; the causes of this bad circulation and stagnation were 'a debility of the brain, from long grief or fear, love, immoderate venery, diseases, the abuse of spirituous liquors, narcotic medicines; a preternatural afflux of blood to the brain, from vehement anger, external cold, suppressions of evacuations of blood, hypochondriacal and sedentary affections; or a siziness [stickiness] of the blood itself, from gross foods, a sedentary life, or other causes'.[4] Melancholy manifested itself in fear and sadness without any obvious cause, and delirium ('reason perverted in thought and words') without fever.[5] Mania seemed closely related to melancholy, and Hoffman later decided that it was a more extreme stage of melancholy.

The great Dutch physician Hermann Boerhaave (1668–1738) also promoted the Newtonian mechanical philosophy in medicine, in which the human body was an hydraulic machine that needed the solid pipes and networks of the body to contain the various fluids. Fluids, thanks partly to older corpuscular ideas and newer mechanical ones, were thought of as composed of particles that might be mixed in different ways, again by mechanical or even chemical principles, although Boerhaave was not an iatrochemist. As with Hoffman, black bile was now to be seen as a kind of sludge, 'thick, black, fat and earthy', that slowed down the proper circulation of the blood.[6]

Melancholy had three stages, the first affecting the whole system via the blood, and inducing various symptoms such as 'a lessen'd appetite, a Leanness: Sorrowfulness; Love of Solitude; all the Affections of the Mind violent and lasting; and Indifference to all other Matters; a Laziness as to Motion'. In an apparent paradox, such individuals would have 'a very great and earnest Application to any sort of Study or Labour' (314).

The second stage was 'an Hypochondriac Disease' or 'The Spleen', and was caused by a further thickening of the black bile and other waste matter in the blood, which was forced into the 'Hypochondriac Vessels' where it stagnated. This area of the body now experienced intensified symptoms, including difficulty breathing and 'a Sense of constant Weight, Anguish, Fulness, chiefly after eating and drinking' (315–16). The last stage of melancholy, a true '*Atra Bilis*', resulted from further corrupt matter being poured into the bloodstream from the existing obstruction and consequent putrefaction in the hypochondria. Now all functions were affected, especially the brain.

Although the influence of mechanical philosophy in medicine cannot be overstated, it was challenged and modified in the second part of the century. Investigations into electricity provided new insights into physical functions and the potential for new therapies. The founder of the Methodist Church, John Wesley (1703–91), actually tested out his electrical devices (see Fig. 8) on his congregation.[7] Albrecht von Haller (1708–77) was a major figure in this shift, with his experiments into the operation of the nervous system yielding the twin concepts of sensibility and irritability, the first of which was to become a word having powerful currency in wider culture. His work in anatomy as professor of medicine at the University of Göttingen brought him a European reputation. Newton's own notion of the 'aether' as a kind of weightless fluid occupying all space further modified mechanism. It was a medium that might vibrate and stretch, allowing forces such as electricity, magnetism, heat, and light to move from one body to another. Electricity, for example, might invigorate the life force, providing new energy and hope to melancholics.

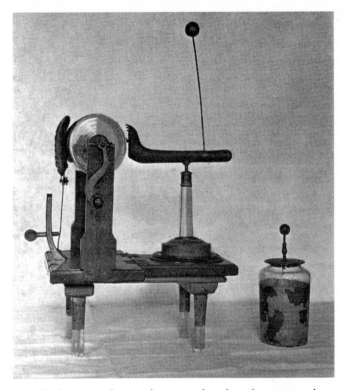

Fig. 8 The discoveries relating to electricity in the eighteenth century gave hope of revivifying the life force of the melancholic. The electrical machine here was designed for the treatment of melancholia by the ever-ingenious founder of Methodism, John Wesley. (*Wellcome Library, London*)

How this affected medical thinking can be seen in Richard Mead's (1673–1754) work. A former student of Archibald Pitcairn and physician to George II, in his later writing he moved away from purely mechanist views of the type we have already seen and towards an idea of the nerves as possibly solid or hollow, but transmitters of a subtle fluid similar to Newton's aether. He wondered whether nerve fluid was electrical in some way, a

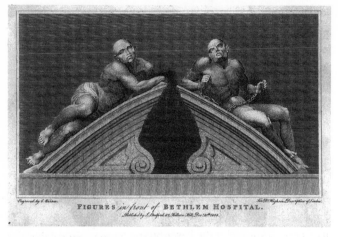

FIGURES *in front of* BETHLEM HOSPITAL.

Fig. 9 Caius Cibber's famous statues of 'raving' and 'melancholy' madness crowned the gates of Bedlam, again reflecting the physiognomical approach to diagnosing depression. Bedlam—the Bethlem Hospital in London—was notorious for being a combination of prison and freak show, 1680. (*Wellcome Library, London*)

speculation that signalled the shift from the vasocentric model of the first part of the century to the neurocentric one of the second. The nervous system, although little understood, was becoming the focus of experimentation via Haller and others.

As to melancholy, Mead stated that changes in the animal spirits or 'active liquor' must be the cause.[8] For him, melancholy was a 'kind of madness' involving sadness, fear, and 'vain imaginations': the other kind of madness was 'furious' of the type illustrated by Cibber's famous statues (see Fig. 9).[9] Passions of the mind could cause problems in the body, 'raising commotions in the blood and humours' (*Medical Precepts,* 77).

The problem of naming melancholic conditions persisted throughout the period. Melancholy as a term smacked of the older, humoral explanations, yet no single new name was forth-coming to replace it. Hence the Spleen stepped forward as a new term that might more accurately describe the process by which melancholy was caused. Physical explanations along mechanical and chemical lines were put forward to explain the malfunctioning of the body that would lead to malfunctioning of the mind, but the names for such related terms jostled with one another. Even by mid-century a literary case of female mel-ancholy conveys the popular and indeed medical problems in defining the malady: in *Amelia*, Henry Fielding's Captain Booth describes the heroine's condition: 'A Disorder very common among the Ladies, and our Physicians have not agreed upon its Name. Some call it the Fever on the Spirits, some a nervous Fever, some the Vapours, and some the Hysterics.'[10]

Nicholas Robinson's article entitled 'Of the Hypp' in the *Gentleman's Magazine* (1732) stated that 'the old distemper call'd *Melancholy* was exchang'd for *Vapours*, and afterwards for the *Hypp*, and at last took up the now current appellation of the *Spleen*, which it still retains'.[11] Even at the time the perhaps con-venient confusion among disease names was noted and duly satirised. Nonetheless, the changing status of melancholy and its depressive cognates was taken seriously by doctors and patients alike, as the correspondence, diaries, and textbooks of the time demonstrate. The vacuum left by Renaissance melancholy may have been filled by a hodge-podge of related terms, but a prodi-gious effort to understand the whirl of depressive symptoms in a new light forged ahead.

The move from mechanism to neurology manifested itself clearly in the displacement of Boerhaave as the main medi-cal authority in the second part of the eighteenth century. The Scot William Cullen (1710–90), a product of the Scottish

Enlightenment and the excellent medical education to be had in that country, rejected hydraulics and explained disease and health through the nervous system. He stated that all diseases are 'nervous' to the extent that they involve the stimulation (under-or over-) of the nerves. The excitation and indeed excitability of the nervous system, especially the brain, might be a cause of melancholy if it malfunctioned: 'if any part of the brain is not excited, or not excitable, that recollection [a memory] cannot properly take place, while at the same time other parts of the brain, more excited and excitable, may give false perceptions, associations, judgements'.[12] Nerve fluid or 'nervous power', again akin to Newton's aether, allowed vibrations rather than flowed through the nerves. This nervous power was prone to either 'excitement' or 'collapse', the balance of which determined health or disease.[13] Cullen had the feel of a vitalist in some parts of his approach, in that the nervous fluid might be seen as an animating and 'vital' force that brought the entire person to life, but in other ways his approach seemed mechanistic, materialistic, and not relying on the ideas of an independent soul proposed by the German physician George Stahl (1659–1734).

In Cullen's nosological system (classifying diseases) melancholia was a 'partial insanity' and caused by 'a degree of torpor in the motion of the nervous power, both with respect to sensation and volition', itself caused by a drier substance in the brain due to lack of nerve fluid. Delirium and mania (into which melancholy might pass) could involve overexcitement of certain areas of the brain, with mental functions overcoming others. One important distinction for Cullen was that hypochondriasis differed from melancholia because dyspepsia (bad digestion) was a strong feature of hypochondriasis, but not of melancholy.

Cullen's treatments for melancholy, like those of the other physicians we have surveyed here, did not veer far from the traditional humoral ones: theories changed, but the approach to cure generally remained conservative and eclectic. What had been used to relieve noxious humours was now getting rid of waste matter in the blood: bloodletting, purges, vomits, and so on, although they came with the warning that they should not be too vigorous for the melancholic's weakened state. Steel and iron tonics might be of use, and their association with spa waters increased. Indeed, towards the end of the century the jovial company and entertainments of the new spa resorts like Bath were thought of as helpful. Good diet and exercise (riding was popular) were perennial favourites, as was the injunction to counter gloomy thoughts by suitable oppositional diversions and the mandatory cheerful surroundings and people.

Dr Johnson's use of the word 'management' in relation to his own melancholy points us towards an increased interest in a more secular psychological approach to managing mental illness in general. No doubt the rise of psychology via Locke in the previous century aided this process, which already had some stimulus in the non-natural of passions of the mind. John Locke (1632–1704), physician and father of modern psychology, had shown that the way ideas were associated in the mind could be perverted, and Laurence Sterne's comic anti-novel *Tristram Shandy* (1759–69) exploited Locke's logic to demonstrate the peculiarities of individual psychology as generated by the inevitable mis-association of words and ideas. More recent contributions by vitalists such as Stahl and Gaub also drew attention to the power of the mind rather than the body to cure melancholy. In tandem with general cultural trends, the second half of the eighteenth century began to regard the life of the emotions as having a significance equal to the life of the reasoning mind, a development that would help constitute the Romantic movement.

The development of Cullen's ideas by Edinburgh physician John Brown (1735–88) was a strong influence on the culture and medicine of Romanticism: his 'Brunonian' theory explained melancholy as a state of under-stimulation or under-excitement (*asthenia;* mania was *sthenic* overstimulation). Brown's promotion of excitability to the heady status of master of human health was appealing because of its simplicity. It enjoyed a great influence in wider culture and literature across Europe and America, even if its shortcomings as a medical theory became apparent within a short period of time, and faded before the end of the Romantic period. Brown's treatment for melancholy was largely alcohol or opium (a stimulant, according to him): no wonder that Brown died an alcoholic and that the poet Samuel Taylor Coleridge (1772–1834) became addicted to opium.

Along with the new medicine of the nerves came the cultural phenomenon of Sensibility and its related concept of the Sentimental. The imagery of the nerves, all delicate, refined, gossamer-light, lent itself to the perception of the sufferer himself or herself as an elevated being, whose physiology reflected moral and intellectual qualities. The contrasting imagery of blocked tubes from bad digestion might be a way of bringing the nervous sufferer back down to earth in a far grosser reality, as the instance of George Cheyne (1671–1743), society doctor and self-proclaimed sufferer of the new-fangled nervous disorders, attests (see Fig. 10). Hailing from Scotland and former student of Pitcairn, he plied his trade in London and Bath, and treated some

Georgius Cheynæus M.D.
di Societatis Regiæ Socius.
Ætat: 59. 1732.

Fig. 10 Dr George Cheyne in 1732, celebrity doctor and self-proclaimed sufferer of the 'English Malady'. He claimed to have cured himself through a strict dietary regime, the 'milk and seed' diet, combined with regular exercise. (*Wellcome Library, London*)

of the most famous figures of his day. His most influential publication was *The English Malady: Or, A Treatise of Nervous Diseases of all Kinds, As Spleen, Vapours, Lowness of Spirits, Hypochondriacal, and Hysterical Distempers &c...With the Author's own Case at large* (1733).[14] Cheyne had a gift for communication with a lay audience, and advocated self-help in terms of diet and regimen. A bad lifestyle led to melancholy ('lowness of spirits') and those cognate disorders mentioned in his title.[15]

Cheyne pointed the finger at the physical causes of mental illness, but also regarded those causes, such as bad diet, as the result of progress and civilisation, a mixed blessing that saw eighteenth-century consumer society advancing in knowledge, sophistication, and wealth, but regressing in terms of its health. In the preface he wrote: 'The title I have chosen for this treatise, is a reproach universally thrown on this Island by Foreigners, and all our neighbours on the continent, by whom nervous distempers, spleen, vapours, and lowness of spirits, are in derision, called the ENGLISH MALADY. And I wish there were not so good grounds for this reflection.' The reason for England being in such a bad state of physical health was clear to Cheyne: 'since our wealth has increased and our navigation has been extended, we have ransacked all the parts of the globe to bring together its whole stock of materials for riot, luxury, and to provoke excess. The tables of the rich and great (and indeed of all ranks who can afford it) are furnished with provisions of delicacy, number and plenty, sufficient to provoke, and even gauge the most large and voluptuous appetites.'[16]

The gorging English were ruining their digestions and their nerves through their very success as a trading empire, and Cheyne did not exempt himself from this moral charge. He claimed that, due to his overeating, 'I swell'd to such an enormous Size, that upon my last Weighing I exceeded thirty two Stone' (Ingram, 89). Although Cheyne reversed his weight by adopting his 'milk and seed' diet, he continued to suffer the ill effects of his previous bad lifestyle: 'I was seiz'd with such a perpetual Sickness, Reaching [retching], Lowness, Watchfulness, Eructation [belching], and Melancholy...that life was no longer supportable to me, and my Misery was almost extreme' (Ingram, 90). Symptoms were protean—physical as well as

86

mental—following the logic of nervous disruption as the period understood it: 'At last, my sufferings were not to be expressed, and I can scarce describe, or reflect on them without Horror. A perpetual Anxiety and Inquietude, no Sleep nor Appetite, a constant Reaching, Gulping, and fruitless Endeavour to pump up Flegm, Wind, or Choler Day and Night: A constant Colick, and an ill taste and savour in my mouth and stomach, that over-came and poisoned every thing I got down; a melancholy fright and Panick, where my Reason was of no use to me: so that I could scarcely bear the Sight of my Patients, or Acquaintances, that had not been daily about me, and yet could not bear being a Moment alone, every Instant expecting the loss of my facul-ties or Life' (Ingram, 90–1). After exacerbating his 'Anxiety and Sinking' (Ingram, 91) by opiates—'a slow poison'—Cheyne found a cure for his melancholy through a vegetarian diet 'so that the stomach need never be cloyed' and abstinence from alcohol.

Cheyne's regimen placed a heavy emphasis on diet, but to aid the free flowing of the alimentary system he also recommended cures that were conveniently similar to those of humoral the-ory. Just as vomits and purges of various types were thought to reduce the excessive quantities of offending humours in the body, so they might 'discharge the Choler, or Bile, and Phlegm from the Liver and Alimentary Ducts, but as by their Successions and Action, they open the Obstructions of that vast Number of Glands situated in the inner side of these ducts (which too are either the cause of, or certainly attend most of the violent Nervous Symptoms) and promote the circulation and perspira-tion' (Ingram, 88).

So, as the eighteenth century progressed, the new idea of melancholy and its cognates became a curious mixture of gross physical disorder stemming from bad digestion and the related but more refined concept of nerves being untuned or 'relaxed'. This odd contrast between the powerful physical and mental sufferings of a Cheyne (for example) and a much more positive discourse of finer-nerved genius requires a little more investigation: how did the Renaissance's melancholic intellectual persist into the age of nerves, spleen, and vapours? What were the alternatives to this cult of melancholic creativity? Where did religious melancholy go? We start answering these questions with an examination of how the new medical theory justified a fashionable notion of melancholy and the spleen.

Fashionable melancholy revisited

The Renaissance discourse of melancholic genius evolved via the medicine of the nerves and spirits into a reworked vision of the depressive—a suffering but inspired person of the eighteenth century. William Stukeley (1687–1785), London physician and sometime antiquarian, in his treatise on the spleen, felt that melancholy most frequently attacked scholars and persons 'most eminent for wit and good sense'. His medical basis for the idea of spleen as wise disease was that 'refined and delicate animal spirits would insure rapid and accurate transmission of sensory data and would provide its possessor with an alert and agile mind. Delicate animal spirits, however, were easily disordered and readily susceptible to melancholic fits.'[17] Sir Richard Blackmore, sometime poet and physician, said that superior wit or genius resulted from a 'peculiar Temperament in the Constitution of the Possessors of it, in which is found a concurrence of regular and exalted Ferments and an Affluence of Animal Spirits refin'd and rectify'd to a greater degree of purity;

whence being endow'd with vivacity, brightness and celerity as well in their Reflections as direct Motions, they become proper Instruments for the sprightly operations of the Mind'.[18] Purity, refinement, vivacity, brightness—not words one would immediately associate with the depressed person, but medical theory had the explanations for such an apparent paradox.

The disease of the Spleen too 'is the particular constitution of the Animal Spirits stimulated, enlivened and refin'd by certain active Principles, to a greater degree than they are in others' (*Essays*, 212). The Spleen, therefore, was 'in reality a generous Constitution, which gives that Acuteness, Vivacity of Imagination, and abundances of Spirits, that exalt the Possessor above the Level of Mankind' (*Essays*, 229–30). Blackmore echoed the Aristotelian idea of balance, even as he replaced the humours with nerves: "Tis therefore a general and just Observation, that those who are endow'd with a moderate Portion of the Spleen in their Complexion, are Persons of superior Sense, and extraordinary Vivacity of Imagination ... '[19] For Blackmore, the animal spirits were crucial in delivering both health and intellect.[20] Melancholy and genius were a consequence of overly refined animal spirits. Too great a quantity of highly refined animal spirits resulted in an excess of wit or spleen, which, in turn, would lead to madness. The Melancholy man and the man of genius 'often approaches to a state of Lunacy'.[21]

Similarly melancholics were thought to possess lively and elastic fibres, those solid parts of the body beyond the nervous system. Fibres in this state would respond to the subtlest sensations quickly and were considered to define the poetic and intellectual constitution. Nicholas Robinson, for example, held that sufferers of the spleen were 'naturally quick of

Apprehension…and of a Mind finely turn'd to contemplate their ideas…'.[22] Little wonder that, as the poet Dryden put it,

> Great wits are sure to madness near allied,
> And thin partitions do their bounds divide.[23]

Hence Aristotle's pronouncements on melancholy were updated by medics for the eighteenth century 'and explained the generation of poets and scholars by the same physiological process which produced madmen and idiots'.[24]

However, medicine was not separate from other discourses: as in previous periods, a male, upper-class elite was the initial beneficiary of such elevated notions, yet the concept of depressive genius was also exploited by other groups in society, including women and people from the lower orders. Medics with a wide popular circulation like George Cheyne and the Swiss Samuel Auguste Tissot (1728–97) argued that melancholy was a result of civilisation and elite refinement. In his essay 'On the disorders of the people of fashion' (1766), Samuel Tissot claimed that fashionable people are subject to melancholy and its cognates because of their modern, 'civilised' lifestyle.[25] The 'best citizens' (male and female) have 'an aversion to simplicity' that causes them to develop 'many diseases unknown to the fields, and which are triumphant in high life' (18). In contrast to 'the labourer', with his coarse nerves, fibres, and balanced perspiration (15–17), the upper sort are feeble in body and mind. Their 'course of life' has 'nothing useful to support it' because they do no meaningful work, so these 'sons of idleness' rely on 'continuous dissipation' 'to defeat the insupportable tediousness of an inactive life' and 'kill time by pleasure' (18). The inevitable consequence of the idle lifestyle is physical and psychological disorder.

It was important for the popularity of melancholy among the fashionable elite that the old equation of mental illness with religion should be removed, or at the very least displaced to other sectors of the populace. There is much debate on the extent to which the eighteenth century was a secularised period, and it is fair to say the degree of religious fervour across the period and within different social groups is varied. As Roy Porter has argued, the detachment of mental disorder from either God or the Devil as a mark of either revelation (as in the Shakespearean wise fool) or punishment for sin removed the moral stigma of nervous disorder. Although there was secular and humanist precedent for valorising genius in the Renaissance, the strong religious element in, for example, Burton's *Anatomy of Melancholy* was an obstacle to the increased popularity, at least in image, of the new melancholy of the nerves of the Enlightenment: 'Freed from contamination by the demoniacal and the vulgar, the elite could luxuriate in the self and toy with mental and emotional singularities, in so far as these squared with other cultural desiderata such as aspirations to artistic genius, refined sensibility, sublimity, or being an "original". Nervous disorders were gentrified and received into good society.'[26] We must remember that melancholy could tip over into madness, however, but this was an extreme end of a very wide spectrum of more or less severe symptoms.

Female constitutions were thought to be finer-nerved than those of men, although towards the end of the century the cult and culture of sensibility meant that men too might swoon, weep, and fall ill due to the powerful impact of the often depressing messages that were being transmitted by their oversensitive nervous systems. The female womb, as in previous centuries, was thought to be a source of disorder that needed regulating, but even the hysteria of Burton's time, a condition thought to proceed from the womb being either under- or over-employed, was by the early eighteenth century regarded as another form of hypochondria. Thomas Sydenham stated that what 'in men we call Hypochondriacal, in women [we call] Hysterical', and that disordered animal spirits were to blame for both genders.[27] It is true that reasons were still invented to distinguish the irrational female from the rational male. The satirist Bernard Mandeville (1670–1733) has one of his characters argue that 'One hour's intense thinking wastes the spirits more in a Woman, than six in a Man.'[28] Such views did not go entirely unchallenged, but were common in popular as well as medical literature. Nevertheless, given the broader cultural confines that entrapped the life choices of women, the depressive symptoms of fashionable melancholy might be potentially enabling.

As in the previous centuries, various women, especially those with the status and money to gain an education, could exploit the possibilities of the older discourses of melancholic creativity as well as the newer ideas about the finer-nerved sensibility that leads to depression. Literary women such as Anne Finch, Lady Mary Wortley Montagu, Elizabeth Carter, and Charlotte Smith have all been described as suffering psychological illness of a depressive nature—Heather Meek uses the rubric of hysteria to characterise these women, but their symptoms of sadness and anxiety bear a strong relation to depression in our time and could equally be termed melancholy in the period.[29] Depression is accompanied by a dizzying variety of physical symptoms that are often the products of psychic stress.

Anne Finch (1661–1720), Countess of Winchilsea, is an early example of a woman of letters in the Enlightenment who recognised both the severity of her condition and the social and cultural advantages that might accrue from it. Her parents died when she was a child, and this seems to be the driver for her consequent lifelong bouts of depression. As a talented poet, Finch expressed her struggle with what she termed 'the Spleen' in poems such as 'On Grief', 'To Sleep', and 'A Fragment Written at Tunbridge Wells', where she was seeking treatment for her condition. Finch's depression was also bound up with political events: she suffered a severe attack when Queen Mary of Modena was deposed, as the Queen had been good friends with the poetess.

Finch's most famous poem, widely anthologised in her period and our own, is 'A Pindaric Ode on the Spleen', originally written in 1694 after Finch's recovery from a serious bout of illness. Most of the poem describes depression in negative terms, and in its various protean forms. Sometimes it will be like a 'dead sea ... /A calm of stupid discontent', and others it will be 'dissolv'd into a panic fear':

> On sleep intruding do'st thy shadows spread,
> Thy gloomy terrors round the silent bed,
> And croud with boding dreams the melancholy head.[30]

Although Finch's poem is largely secular, she invokes the Christian myth of sin and death allowing disease—physical and mental—into the world, so that depression becomes part of a Fall from an ideal harmony of body and mind into a representative of the disorder brought about by Satan. At another point in the poem she also engages with religion, 'that should enlighten here below', but instead is brought by Spleen into 'darkness and perplext/with anxious doubt, with endless scruples vext'.

Finch was not the only person to represent the range of possible melancholic states, and it was clear that poets were often making a conscious decision to represent the milder and potentially more creatively balanced forms of depression. Thomas

Gray (1716–71), author of one of the most famous melancholic poems in English literature, the 'Elegy written in a Country Churchyard', clearly outlined two kinds of melancholy:

> Mine, you are to know, is a white Melancholy, or rather Leucocholy for the most part; which though it seldom laughs or dances, nor even amounts to what one calls Joy or Pleasure, yet is a good easy sort of a state, and ça ne laissé que de s'amuser... The only fault of it is insipidity; which is apt now and then to give a sort of Ennui... But there is another sort, black indeed, which I have now and then felt,... it believes, nay, is sure of every thing that is unlikely, so it be but frightful; and, on the other hand, excludes and shuts its eyes to the most possible hopes, and every thing that is pleasurable; from this the Lord deliver us! for none but he and sunshiny weather can do it.[31]

Gray gives us 'white' melancholy (leuchocholy) versus 'black' melancholy (which invokes the old black bile and the darkness of true depression): one is a 'good easy sort of a state' and borders on the pleasurable; the other is a condition of despair, or rather the impossibility of hope and faith. The coexistence of the two forms is a feature of this century and others, and it is clear from Gray's correspondence that he suffered attacks of the 'black' variety in which despair and anxiety threatened his sense of himself as a rational being.

Melancholy and society

Fashionable melancholy may have taken much of the representational attention throughout the century, but the sheer misery of the condition was still present and itself demanded the consideration of serious thinkers, including the total defeat of hope in suicide (although even that phenomenon managed to gain at least a theoretical popularity later in the century).

Different social groups suffered a variety of pressures that led to melancholy, some of which were recognised at the time, and some that were not. The restrictions on women, for example, are described by Mary Wollstonecraft's writing at the end of the century, and indeed in her own depression. We have already seen Anne Finch's problems with female authorship, and Mary Jones (1707–78), poet and acquaintance of Dr Johnson, wrote as 'a relief from madness' and depicted the lot of women in a despairing manner in her poetry. Sarah Scott (1720–95), novelist and author of *Millenium Hall* also portrayed melancholy in her novels and linked it to the circumstances of women. Scott similarly empathised with slaves, a group that she and Wollstonecraft saw as parallel with women in oppression by white men. In Scott's *History of Sir George Ellison* the 'depression of spirits' of slaves dissipates when they are helped to freedom.[32] Theodore Parsons, in his *A forensic dispute on the legality of enslaving the Africans,* argued that the spirits of slaves are understandably depressed and that this degradation and melancholy also affect their ability to participate in any form of education. What might be regarded as a natural dullness or stupidity is actually depression.[33] Observations like this might seem obvious to our age, but the eighteenth century was only slowly developing a sociological imagination, and persisted in a tendency to blame personal moral deficiencies for either crime or insanity (as well as praising personal qualities for good and sane behaviour). It should be noted that the picture is not always one of men causing mental illness in women: Judith Madan (née Cowper, 1702–81), was a poet advised by the great poet Alexander Pope to write to counter her recurring depression. She was also the aunt of the famous religious melancholic poet William Cowper, and a possible influence (genetic or otherwise) on her nephew.

As today, financial matters could bring on depression. In the personal realm we have the famous instance of Georgiana Cavendish, Duchess of Devonshire (1757–1806), whose high living included massive gambling debts. The eighteenth-century novel is full of characters who fall into melancholy and disease because of family debt: men were often responsible for the family estate in the upper classes, but women too could be afflicted with depression, especially if they had been left without the monetary and social protection of a husband.[34] Similarly, in reality and in representation, early macro-economic disasters like the 'South Sea Bubble' in 1720 brought melancholy and even suicide, as the title of one publication by Sir John Midriff MD asserted: *Observations on the Spleen and Vapours: Containing remarkable cases of person of both sexes and all ranks from the aspiring directors to the humble bubbler who have been miserably afflicted with these melancholy disorders since the fall of the South Sea and other public stocks.*[35] Midriff claimed that this eighteenth-century stock-market crash had brought about an increase in depressive patients in his practice, and went on to describe a variety of cases. The writer Eustace Budgell (1686–1737) committed suicide as the result of his debts from the Bubble, as well as allegations of forgery. Famously, he jumped from a boat at London Bridge in 1737 and left a suicide note saying 'What Cato did, and Addison approved, cannot be wrong.'[36]

Suicide was generally a deeply unfashionable aspect of melancholy. The taking of one's own life in the despair induced by depression was partly thought to be of a piece with the English Malady: a function of the melancholy-inducing British climate. Notable depressive suicides included John 'Estimate'

Brown (1715–66), a Newcastle author, moralist, and poet. Stephen Duck (1705–56), the so-called 'thresher' poet, rose to the eminence of a reverend in defiance of class conventions, but the pressure of that rise no doubt contributed to his eventual depression and suicide. Although the increasing secularisation of the eighteenth century brought about a change of attitudes—the depressive philosopher David Hume's (1711–76) 'Essay I, On Suicide' argued against the traditional Christian idea of suicide as a crime against God—this shift took a long time and was more significant in non-traditional sectors of society.

Suicide could be fashionable if it provided proof of one's finer-nerved superior sensibility, and this became a more popular notion in the second half of the eighteenth century and on into the Romantic period. Nor was it merely—and contrary to the myth—a purely English Malady: suicide was a fad elsewhere. Lady Mary Wortley Montagu reported its vogue in mid-century Italy. Of course, the fictional self-murder of Goethe's sensitive, creative, frustrated lover *Young Werther* in his famous novel (*The Sorrows of...*, 1774) sparked at least a theoretical shift to fashionable suicide, although debate continues as to how much difference Goethe's novel really made to European and American suicide rates. In the British context, the spectacular representation of the (alleged) suicide of Thomas Chatterton (1752–70) painted a picture of the cruel world crushing the spirits of a promising but sensitive young poet.

Romantic melancholy

Ay, in the very temple of Delight
Veil'd Melancholy has her sovran shrine.
(John Keats, 'Ode on Melancholy')

With Werther and Chatterton and their associated mythologies, we arrive firmly in the Romantic period. The idea of a creative or at least pleasurable form of depression continued apace, encouraged by the evolving influence of Brunonian (John Brown's) medicine and the related concept of sensibility. Poets, writers, and artists trembled with sensibility, were overexcited in their supply of nervous life force, and exhausted their energies: they lived fast and died young (literally as well as theoretically in the case of Keats and Shelley). The increased focus on the self, as opposed to the civic and social responsibilities of the earlier period, meant that Romantic melancholy intensified the idea of this disease as a marker of self-reflection, permitting contemplation of the deepest aspects of existence, and generally typifying creative genius.

Charlotte Smith (1749–1806) was a female poet who suffered from depression and who used it for her poetry. She had a well-to-do country childhood, but regarded her early marriage as a form of slavery and proceeded to suffer the deaths of several of her children, including her favourite, Augusta. These events, understandably enough, plunged her into depression and anxiety. In her *Conversations Introducing Poetry* (1804), a children's book, she baldly stated—via the character of a teacher talking to a child—the general opinion of the literary sort on the relationship of melancholy to genius: 'It has been observed, George, that almost all men of genius have a disposition to indulge melancholy and gloomy ideas; and in reading our most celebrated poets, we have evidence that it is so.'[37] Smith, via the *Elegaic Sonnets* as well as her prefaces to various works, presented herself as a true melancholic and therefore genuine poet, and attempted to forge a space for herself in the tradition of melancholic sensibility. Thus in her first sonnet in the volume she paints herself as both cursed and blessed by her gift: 'those paint sorrow best who feel it most'.

At the level of representations at least, Romanticism was hospitable to a certain version of melancholy, a version that was conveniently mild. John Keats's 'Ode on Melancholy' and Samuel Taylor Coleridge's 'Dejection: An Ode' are markers of the Romantic valorisation of a pleasurable sadness, an excuse for meditative midnight excursions into the creative psyche, as well as a psychological space for envisioning social change. The poets in question suffered badly themselves from melancholia and associated physical problems—Keats's 'nerves' were partly a product of his consumptive illness—but they shared the ability to portray melancholy as distinguishing the mind of the suffering genius. The Romantic period was the high point of the cult and culture of melancholy, but developments in the following century were to pave the way for modern depression.

IV

—∞∞∞—

VICTORIANS, MELANCHOLIA, AND NEURASTHENIA

Yes, here and there some weary wanderer
In that same city of tremendous night,
Will understand the speech and feel a stir
Of fellowship in all-disastrous fight;
'I suffer mute and lonely, yet another
Uplifts his voice to let me know a brother
Travels the same wild paths though out of sight.'

James Thomson, 'The City Of Dreadful Night', 1874

The word melancholia, consecrated in popular language to describe the habitual state of sadness affecting some individuals should be left to poets and moralists whose loose expression is not subject to the strictures of medical terminology.

Esquirol, 1820[1]

James Thomson's 'The City of Dreadful Night' marks the Victorian period as one in which depression became part of the urban landscape, an industrialised nightmare in which God was absent and despair was almost inevitable. Despite his intent, the poem scarcely gives one hope of therapeutic release, ending as it does with the statue of *Melencolia*, which itself is a reworking of Dürer's famous figure from the Renaissance. This was also the century in which 'melancholy' was moving towards 'depression', a shift partly enabled by the advent of psychology as a profession, increasing knowledge about brain and nerve anatomy, and the overall move from the idea of melancholy as a disorder of the intellect to the notion of melancholia and depression as a mood or 'affective' disorder.[2] 'Depressing' passions in themselves could slow the circulation of the blood and lead to faults in the brain. This swing from intellect to emotion as a basis for depression has largely persisted to the present day, despite some notable exceptions.

Psychiatrists in Europe and America strove to wrest the definition of mental disease from the realm of the popular and to prove that their profession could attain the status of a science. New mental hospitals meant that this was the age of the 'birth of the asylum', and a time in which clinical and anatomical investigation would reveal the mysteries of the human mind—supposedly. The incremental movement of the concepts of both melancholy and mania via this new faculty of psychology was to culminate in the early twentieth century in Kraepelin's influential theory of manic depression that would encompass most mood disorders, including what we would now recognise as major depression. Throughout the nineteenth century the free communication between European and American alienists (or 'mad doctors') meant that they largely shared the notion that melancholia was a product of 'brain disease', was inherited in predisposition, could be combined with mania, and was primarily an affective disorder (see Fig. 11).[3] Earlier in the century the blood supply to the brain was a central medical focus in the quest to find an explanation for mental problems, but

Fig. 11 'Melancholy passing into mania' (1858) by W. Bagg after a photograph by Hugh Welch Diamond (1809–86), amateur photographer and physician at the Springfield Hospital in Surrey. Photography was thought to be a useful new tool with which to record the psychological state of the patient with greater scientific accuracy. (*Wellcome Library, London*)

increasingly the developments in ideas about the nerves took over in the second half of the century. Both jostled alongside each other for the entire period, but with different emphases at different stages. Insofar as the blood supply to the brain affected the nervous system, the vasocentric (blood) and neurocentric (nerves) theories had common ground.

The status and nomenclature of depression also remained unstable in the period, although there were elements of continuity with the previous century, such as the persistence of 'nerve theory' and 'sensibility' and medical and social movements that continued to define the melancholic individual. Cullen's introduction of the term 'neurosis' in the eighteenth century marked the importance of the nervous system in causing physical effects that would impact on the functioning of the mind, and it was not until the influence of Freudian psychoanalysis at the end of the nineteenth century and—in the UK—after the First World War, that it would acquire its meaning as a purely mental concept. Before then, no one was sure how to separate a problem with the nervous system from a 'pure' disease of the mind, and psychiatrists tended to leave the matter open until more substantial proof was forthcoming. Daniel Noble, a doctor practising in Manchester in the middle of the century, stated that 'practically, it is always difficult to draw the boundary line between what are commonly considered purely Nervous maladies, and diseases of the Mind, on account of the connection subsisting amongst all the nervous centres and the correlated physical states'.[4] Even the 'new' disease of neurasthenia at the end of the century does not look so very innovative when one considers it in the light of theories of nervous over-excitation and depletion of vital energy promulgated in the Romantic period and before by the likes of Cullen and John Brown.

Later nineteenth-and early twentieth-century ideas of 'nerve force' defined the concept of the depressive, but the terminology was unclear for most of the period. There was a specific medical literature on melancholia, as we will see in more detail, but there was also a more generalised phenomenon of 'depression of spirits' and nervous exhaustion, which might stem from nervous sensibility or temperament (a form of hangover from humoral theory reinvigorated by Cullen), that did not necessarily fall under the rubric of melancholia. When 'neurasthenia' arrived, it was such an all-inclusive concept that it became popular precisely for the reason that it gave a name and a medical explanation for a huge range of symptoms. As Arthur Benson put it even at the end of the Victorian period: 'Neurasthenia, hypochondria, melancholia—hideous names for hideous things—it was these, or one of these. The symptoms a persistent sleeplessness, a perpetual dejection, amounting at times to an intolerable mental anguish. The mind perfectly unclouded and absolutely hopeless.'[5] He had suffered a period of depression around 1907– 09 and, even then, the nomenclature was inadequate to describe his state. Like the 'hard' and 'soft' meanings of melancholy—or black and white as Thomas Gray called them in the eighteenth century—defining depression was a difficult task throughout the period, and national differences as well as medical fashions added to the complications.

In the last chapter we left the Romantic popular and literary culture of melancholy and sensibility, a climate in which attitudes to mental illness of this type could be positive, or at least linked with the type of the artist: Romantic heroes were supposed to be 'mad, bad, and dangerous to know', as the much-maligned Lady Caroline Lamb described Lord Byron. Some vestiges of this movement persisted in the artistic and literary discourses of the Victorian age, such as the pre-Raphaelite glo-rification of the supposedly sickly sensibility of a Keats and the sexualised but consumptive and depressive 'stunners' who pop-ulated their paintings. The social environment of the Victorians generally discouraged notions of idle poetic melancholy and encouraged a new evangelical spirit of personal self-discipline, a work ethic—bolstered by the Industrial Revolution—that did not tolerate unproductive labour. Meanwhile the psychiat-ric movement argued, at least in part, that nervous exhaustion affects all classes, and, via new advances in physiology, demon-strated a significantly improved knowledge of the actual work-ings of the nervous system—including reaching a consensus before the end of the century that electrical impulses were the mechanism for the transmission of signals from the nerves to the brain. The mystique of melancholy became less tenable in this new Victorian environment.

New technologies also meant new metaphors for the exhaus-tion of the vital energy or nerve force. The dominant metaphor for nervous breakdown was economic, with much talk of over-spending one's finite reserves of energy, potential bankruptcy brought on by a profligate (or too dedicated) lifestyle.[6] The bur-geoning industrial capitalism of the age naturally lent itself to such notions of the microcosm of the individual reflecting the general economy. Other common metaphors for the nerves and their energy supply included the obvious mechanical ones of batteries and wires, steam engines, or furnaces. Fine cloth might also be compared to the fabric of the nerve networks and the brain's grey matter. The old metaphors of the body as a machine, therefore, were updated for the new age and its new technologies and social differences. Women, of course, were considered to be more delicate machines, or less blessed with nerve force, than men, and were treated accordingly—a convenient reinforcement of Victorian social roles.

Clearly some aspects of the metaphorics of nerve force were gendered, although the basic idea of the nervous system was genderless in the sense that both sexes possessed a nervous system, and the chaos at the end of the eighteenth century in terms of the fashionability of feminised men of sensibility and 'feeling' demonstrates the ability that the new model of the nerves had to unsettle traditional conceptions of the gendered body and mind. However, there were still ways of inflecting a supposed greater delicacy and susceptibility to diseases of the nervous system and psychological problems on women to support broader discourses of gender difference and the (subservient) role of women in Victorian society.[7] Women's wombs were a source of disruption—an old theme, but reworked for a new age. In the realm of medical diagnosis and treatment, however, class was an important factor that interacted with gender. When Virginia Woolf suffered a breakdown including threats of suicide and bizarre hallucinations, she was treated as a neurasthenic rather than being confined to an asylum, as would have occurred had she been male and working class. Woolf has been held up as a icon of victimised womanhood but, wrong-headed though the treatment that Woolf received might have been, to be in a different class would have been worse.

The general message—for both genders—was that 'nerve force' needed to be carefully managed, otherwise it would be destroyed, and with it the mental health of the individual. This management resonated with the ideas on regimen that had been common since the recommendations of the ancients: too much drink, sexual activity (especially masturbation in men), late hours, bad company—all could drain the body of vital force or fluid. There was a flip-side for the industrious Victorian, however: overwork, whether it be study, literary and artistic creativity, or application to one's business, might also push the nervous system into depression, a lowering of nervous energy. Men were thought to be susceptible to this factor, whereas women were controlled by the reproductive system and derived their mental ailments from that clichéd source. Depression might come from opposite moral directions, then: the ne'er do well or the model citizen whose only crime against health was to try too hard. Homosexual activity in men was also thought to be a waste of vital energy, although it is clear from the case of the poet John Addington Symonds that his struggle with the fact of his own homosexuality, in a society that almost universally condemned it, led directly to his own depression.

It is hardly surprising the Victorians had the sense that the rapid pace of change—in all areas of life seemingly—was very likely to result in depression unless carefully managed. The rather gloomy second law of thermodynamics stated that the total amount of energy in the universe was gradually decreasing, and it was easy to make an analogy with the individual's store of nerve force, even if the doctor's treatment was predicated on restoring energy to the depleted person. Matthew Arnold expressed the spirit of the age:

> For what wears out the life of mortal men?
> 'Tis that from change to change their being rolls;
> 'Tis that repeated shocks, again, again,
> Exhaust the energy of strongest souls.
>
> ('The Scholar Gypsy', l. 141)

Again, men were most prone to these 'shocks' because they were 'more exposed to numerous sources of cerebral excitement in the worry and turmoil of the world'.[8] Relentless change in all spheres of life was bound to have an effect, even on those with the greatest supply of nerve force.

The traditional support of religion (which we have seen in previous periods could be a cause of deep depression as well as a comfort) was also being assaulted by Darwinism and Marxism, themselves in part developments of Enlightenment atheism. Victorian literature and art depicts the crises of faith that many people were experiencing in real life. Tennyson's poem 'In Memoriam' ostensibly mourned the death of his friend Arthur Hallam, but moved through the stages of loss, grief, and religious doubt beautifully at epic length. Tennyson resolved his poem in the affirmative as far as Christianity was concerned, but the enormous popularity of the poem derived as much from the depiction of his emotional despair and depression as it did from the 'rescue' of Tennyson's faith at its conclusion.

Melancholia and the physicians

The theories of melancholia and the various nervous disorders associated with a depressed nerve force, and consequently depressed spirits, followed a variety of routes throughout the century, leading up to the consolidation of Kraepelin's definition in the early twentieth century. Treatment or case management were equally heterogeneous. Psychiatry was an international phenomenon and different cultures contributed at different times to the ongoing progress of knowledge—assumed or real—about depression. The French, for example, had a more centralised system that encouraged their doctors of the mind to respond to changes in the political regime, whereas British psychiatry was exercised by the long and powerful influence of evangelical Christianity, and felt the pressure (as illustrated by Tennyson and Arnold) to reconcile their claims to scientific knowledge with the need to affirm the existence of God and the immortal soul. The Germans were less concerned about the difference between organic 'brain disease' and purely mental illness than the British. Later in the period, American psychiatry was to prove more open to Freudian psychoanalysis, as indeed was the whole society. In the following parts of this chapter, we will shuttle between different nations and cultures as we trace the often uneven development of ideas about melancholia and related depressive states through the century.

The eighteenth century has not had a good press when it comes to the treatment of the insane in asylums: the image of the Bethlem Hospital in London—'Bedlam'—making a profitable show of its patients is enduring. But there were other, more humanitarian shifts afoot that advocated an approach that would feed into the nineteenth century. At the Quaker York Retreat, Samuel Tuke argued that 'In regard to melancholics, conversation on the subject of their despondency is found to be highly injudicious. The very opposite method is pursued. Every means is taken to seduce the mind from its favourite but unhappy musings, by bodily exercise, walks, conversation, reading, and other innocent recreations.'[9] This method of distraction, which sounds very modern in one way, was accompanied by a more enlightened attitude to mental illness in general, in which the popular method of scaring the patients into submission to the asylum regime was replaced by kindness and persuasion where possible. Tuke pointed out that 'the terrific [i.e. terror-inspiring] system of management' would actually 'fix for life, the misery of a large majority of the melancholics'.[10] It has to be observed that Tuke was not in the majority, however, and that many asylums still had recourse to harsher regimes predicated on older views of the insane as somehow bestial.

Philippe Pinel (1745–1826) advocated a similar 'moral management' or non-medical treatment of lunatics and protested against the 'rigorous system of coercion' apart from the potential suicides, and wished instead to divert the attention of melancholics away from their obsessive thoughts.[11] Pinel became the Physician in Chief to the Paris mental asylums after the French Revolution and was said to have freed the mad from their chains. His transformation of asylum practice in the next decades made him something of a hero to those advocating a revolution in psychiatry as well as politics. Such a humanitarian approach did not necessarily sit well with psychiatrists as it seemed to suggest that a lay person could supply all the kindness necessary, but they soon learned to combine aspects of moral management with their existing therapeutic armoury (letting blood, water treatments, etc.). Pinel defined melancholia in the traditional sense of a partial form of insanity with a limited number of delusions, perhaps only one. Melancholic symptoms were 'taciturnity, a thoughtful pensive air, gloomy suspicions, and a love of solitude' (*Treatise*, 136). In this attitude Pinel was reflecting the Enlightenment position that melancholics were sane other than their peculiar objects of fixation: their thoughts were at odds with reality. Pinel was of his moment in that, as well as identifying various psycho-sociological causes such as 'ungovernable or disappointed ambition, religious fanaticism, profound chagrin and unfortunate love', he also found 'events connected with the Revolution' to be stimulants to depression (*Treatise*, 113). Melancholia was not a generalised 'disease' in the modern sense. Each person's physical and psychological constitution—especially nervous—predisposed them to an individual outcome, much like the humoral theory of previous centuries. No virus or gene or bacterium was the 'cause' of melancholia, only the infinite variety of individual nervous constitutions and the life events that were visited upon them.

Pinel's favourite pupil, Jean-Etienne-Dominique Esquirol (1772–1840), was enormously influential in promoting the need for an appropriately professional setting for the cure of the melancholy, including properly trained psychiatrists in hospitals specially built for mental patients. He also stressed the need to roll out the advance made in avant-garde Paris across the whole nation. Esquirol broke from the teachings of Pinel when he pronounced that the old terminology of melancholy would no longer do for the brave new world of faculty psychology. For him, melancholy was now a disorder of the emotions, not intellect, and should be called 'lypemania', 'a cerebral malady, characterised by partial, chronic delirium, without fever, and sustained by a passion of a sad, debilitating or oppressive character'.[12] Lypemania was a form of 'monomania', a partial insanity ('gay or sad') focused on one object.[13] In terms of its content, it still seemed very much like the old melancholy, and still had 'illusions of the senses, and hallucinations' (*Mental Maladies*, 205). Objectless fear was also a very strong element in the equation. Melancholia did not include mania here, however, and he also separated hypochondria from melancholia, as hypochondria had no delirium and involved an exaggeration of sufferings (*Mental Maladies*, 203). Esquirol's move in cutting out the group of monodelusional states with an exalted mood had the helpful effect of restricting melancholia to the lowering disturbances, a definition that tended to persist until the end of the century.

Fig. 12 'A man whose face exemplifies the melancholy temperament'. This drawing was for the first English version of Johann Caspar Lavater's *Essays on physiognomy* (1789). (*Wellcome Library, London*)

It had been a long tradition in medicine and art that the mind could be read in the face (see Fig. 12). Esquirol invoked the fashionable pseudo-science of physiognomy to describe the mind of the lypemaniac via the physical features: 'lean and slender, his hair is black, and the hue of his countenance pale and sallow... The physiognomy is fixed and changeless; but the muscles of the face are in a state of convulsive tension, and express sadness, fear or terror; the eyes are motionless, and directed either towards the earth or to some distant point, and the look is askance, uneasy and suspicious' (*Mental Maladies*, 203). See Fig. 12.[14]

The list of causes invoked by Esquirol were not confined to lypemania, and reflected the complex causation described by authors at least as far back as Burton's *Anatomy*. Emotional disruptions ranked high as usual, also hereditary predispositions, certain climates and idle or sedentary occupations were hotbeds of depression. Financial problems, an immoral lifestyle, and other diseases—like consumption—seemed to be a problem, while key life events and stages could trigger the illness, such as childbirth and thwarted love. It was not clear whether women suffered more than men. Esquirol too considered 'moral medicine' to be the ideal form of treatment: he stated that it 'seeks in the heart for the cause of the evil...sympathizes and weeps...consoles' and 'revives hope', not unlike a sentimental novel with its improving message of sympathy for the suffering person (*Mental Maladies*, 226). Other treatments were reminiscent of the traditional ones: diet, exercise, travel, calm distractions, and notably avoidance of a damp climate in favour of somewhere 'dry and temperate' with 'a clear sky'—not a recommendation for the legendarily splenetic clime of Britain. As Keats had put it: 'Who would live in a region of mists...when there is such a place as Italy? It is said this England from its clime produces a spleen, able to engender the finest sentiments, and cover the whole face of the isle with Green—so it ought, I'm sure.'[15] Nevertheless, the rapid growth of British health resorts, whether spa towns or seaside destinations, demonstrated an optimism that specific native airs might be good enough to cure all kinds of conditions.

'Lypemania' persisted in French psychiatry for a time, but did not take hold in Britain or Germany. It was indicative of the shift from melancholy to depression, but did not satisfy the new demands of the changing social and medical conditions elsewhere in the nineteenth century.

Although our focus in this book is on depression rather than mania, the history of the two are bound together and, in the mid-nineteenth century, were more formally theorised by two students of Esquirol, and would lead to the concept of manic depression so prevalent in the following centuries. In 1854 Jules Baillarger (1809–90) coined the new disease, *la folie à double forme*, that had regular intervals of melancholia and mania ('excitement'). Two weeks later Jean-Pierre Falret (1794–1870) claimed his own idea of *la folie circulaire* which—he trumpeted—he had been developing with full knowledge of his students for the last ten years, having published on the subject—although not using the name—in 1851. There followed a tangled argument about who could claim the glory for describing this 'special type of insanity' (Baillarger), but the combination of their ideas influenced thinking on depression for the rest of the century and beyond.

Other nations were not idle in their attempts to bring melancholia into line with the brave new world of psychiatry. The German physician and Professor at Leipzig, Johann Christian Heinroth (1773–1843), produced an important textbook dealing with 'disturbances of the soul' that echoed Brunonian ideas of under- and over-excitation, and called depressions *veseniae asthenicae*, involving disturbances of the 'disposition', 'spirit', or 'will'—one of which was melancholia. His holistic notion of the soul's interaction with the body led him to coin the term 'psychosomatic'. He defined *melancholia simplex* (pure) as

> Paralysis of the disposition, that is, loss of freedom of the disposition accompanied by depression, withdrawal into oneself, and brooding over some loss, death, pain, or despair.

Restless, anxious, rapid movements, or a fixed stare. The patient is insensitive to everything except the interests of the fettered disposition; he sighs, he weeps, and laments.[16]

Heinroth pointed towards the shift to the nineteenth-century perception of melancholia as a disorder of the emotions rather than the intellect by arguing that the '*idée fixe*' or single, fixed idea on which the deluded melancholic would focus was a consequence of the disposition being 'seized by some depressing passion, and then has to follow it, and since this passion then becomes the dominating element, the intellect is forced by the disposition to retain certain ideas and concepts'.[17]

Heinroth saw other types of melancholia as more extreme forms of the condition, and also mentioned religious melancholia as a 'subspecies' of the illness, a problem that he saw as entirely divorced from any actual contact with supernatural forces and a delusion on the part of an overly religious person. Religious melancholy was to die out as a category as the process of secularisation continued through the century and into the next, but it was clearly a significant problem during that period.[18] His description of melancholia smacks of the Romantic moment: 'in melancholia the disposition has lost its world, and becomes an empty, hollow Ego which gnaws at itself'; by contrast, the insane disposition is 'torn and removed from itself' (*Textbook*, 221). Melancholia could be hereditary, but Heinroth emphasised the psychological causes of the illness.

In keeping with a more psychological approach, Heinroth advised that 'friendly sympathy and persuasion' should be used to find the 'sources of the disease, such as a great loss or the fear of such a loss; and then, if possible, by deflection of these sources'. If this early intervention did not prevent a worsening of the problem, then the patient needed to be removed from the situation, 'forcibly' if necessary, and 'some new interest must be awakened', possibly by travel ('a universal medicine'), and with 'much excitement, much discomfort, and much activity' (ii:358). All this was designed to jolt patients out of their self-absorption: if things became worse then other physical remedies or techniques might be used, including the use of drugs and a 'swing machine' to counter apathy—or even a straitjacket if there was extreme overexcitement. Although he advocated a sympathetic approach, Heinroth balanced it with the need to protect both the patient and carers. Idleness and solitariness were to be avoided, in the best traditions of melancholy cure: even purges might be used. Heinroth emphasised that due regard for the individual character of the patient must be exercised for effective treatment, and that patients should be encouraged by any means possible when they were on the track to improvement.

Baron Ernst von Feuchtersleben (1806–49) was an Austrian physician—dean of the faculty of medicine in Vienna—and poet/philosopher. He continued the notion of melancholia—a term he thought too restricted—as partial insanity, concentrating on a fixed delusion as the key to the condition. It mattered not what the actual object of fixation was, more the mechanism by which it came to dominate the individual's mental life.[19] Again, beyond the delusion the afflicted person could act in a perfectly rational manner. Feuchtersleben considered that the particular delusion that life 'must be quitted'—either through fear of death or weariness of life—corresponded with the more narrow meaning of melancholy and that this was like the lypemania of Esquirol. In physical terms, melancholy meant that 'the nervous vitality languishes at its root, and the vitality of the blood, deprived of this stimulant, is languid in all its functions' (135). The patient's body therefore manifested externally this inward sluggishness, with slow breathing, weak pulse, shrivelled skin, paleness, and constipation. The causes of this state of fixed delusion were defined largely in physical terms: there might be a problem in the brain itself, or—and this had a censorious moral ring to it that was to continue throughout the period—'sexual excesses' might have taken their toll on the nervous system.

Even bodily sensations might lead on to delusions and hallucinations. Stomach and abdominal problems could also cause fixed delusions for melancholics, and needed to be treated accordingly. If the patient was suicidal, 'sense of honour, duty and religion, are perhaps the only interests which can rouse the deadened vitality of the mind' (348). Judicious application of emotional stimulation, even fear, could assist in jogging a patient into escaping a depressive mood. The German tradition had a certain latitude in the use of apparently harsh methods in the service of the greater good: cure. Again, due attention to the individual's particular psychical and physical make-up was necessary for effective treatment.

The influential German psychiatrist Wilhelm Griesinger's (1817–68) *Mental Pathology and Therapeutics* (1867, German first edition 1845) provided the refrain for the 'brain psychiatry' of the second half of the nineteenth century and the bio-medicine of the twentieth: 'Pathology proves as clearly as physiology that the brain alone can be the seat of normal and abnormal mental action; that the normal state of the mental process depends on the integrity of this organ.'[20] All mental disease was only a symptom of the malfunctioning brain, according to this view: 'we must not speak of diseases of the soul alone...but of disease of the brain'. Although this can be, and has been, taken as a crude reduction of human mental illness to physical dysfunction,

Griesinger was more subtle and uncertain in his speculations than these statements suggest.

As with Heinroth, he stressed the need to consider individual circumstances very carefully, as often many factors were involved in mental illness. Hereditary predisposition was important, and took into account Morel's theory of degeneration—the idea that some social groups were becoming weaker through the generations, later given further resonance and impetus by Darwin's theory of evolution. Degeneration could occur through a kind of reverse evolution, in which man was in danger of regressing back to the primitive if he did not moderate the abuses of the modern lifestyle. The psychiatry of degeneration, represented in Britain by Theo Hyslop (1864–1933), was another spoke in the wheel of the continuing association of artistic creativity with melancholy. This movement ran contrary to the Romantic legacy of celebrating artistic madness and melancholia by arguing that artists were in fact instances of the backwards evolution of at least certain parts of the human race: morally deviant, psychotic, and physically feeble. Artists were not to be celebrated as suffering heroes, but condemned as degenerate undesirables.[21]

Although multiple physical factors might impact upon the functioning of the brain, Griesinger also accepted the traditional line that 'the depressing emotions, when long continued, grief or anxiety' were factors that could arrest the circulation of the blood and thus have physical consequences. Emotions of a more sudden or intense variety could also produce 'a state of intense irritation of the brain', and could impact on other organs that might in turn affect the brain and thence lead to mental illness (167). Griesinger therefore continued the stress of previous eras on the interaction of mind and body, but provided a modern rationale for that impact. He too did not shy away from moral judgements about mental disease, citing masturbation, excessive sexual activity, and drunkenness as 'mixed' causes. The majority of cases were not of this nature, however, and could have physical sources as disparate as head injuries and pregnancy.

For Griesinger, all mental illness was part of one disease process, with each condition constituting a different stage on a continuum. This concept of unitary psychosis echoed the eighteenth-century notions of Cullen and Brown in some respects, and developed throughout the nineteenth century, with consequences for the later definition of depression, as we shall see when we come to Kraepelin. Griesinger discussed 'States of Mental Depression' and considered the melancholic stage—'a state of profound emotional perversion, of a depressing and sorrowful character'—as the first move towards more serious states. This melancholic state was different from normal sorrow or grief 'by its excessive degree, by its more than ordinary protraction, by its becoming more and more independent of external influences' (210). 'Mental Depression' was soon to lose the 'mental', but was still adjectival at this point, and not a disease entity in itself.

Following on from the melancholic stage was hypochondriasis, 'the mildest, most moderate form of insanity', in which 'the emotional depression proceeds from a strong *feeling* of *bodily illness* which constantly keeps the attention of the patient concentrated upon itself'. Again, the intellect was unimpaired, and the emotion of fear 'justified by reasons which are still within the bounds of possibility'. The patient would have real sensations in the body, but they were not based in any physical problem. If the hypochondria worsened, it could pass into melancholia proper, a state of 'mental pain' which 'consists in a profound feeling of *ill-being*, of inability to do anything, of suppression of the physical powers, of depression and sadness, and of total abasement of self-consciousness' (223).

A particular concern for this age was the problem of the will, a moral concept that implied the power of the individual to control him or herself to be a reflection of literal moral fibre: here Griesinger points to the patient's lack of decision and productivity or activity as a sign of a disorder of the will. Women and children were thought to be less mature in their will than men, and women might spread their nervous disorders down the generations, thus imperilling the race in general. As well as these symptoms the sufferer might become more unhappy, irritable, anti-social, and self-absorbed. The melancholic would suffer chronically, sometimes with remissions, and would also be physically inactive.

Sometimes melancholia could worsen into mania, a condition that involved greater 'persistent excitement and exaltation of the will' than melancholia, being a 'State of Mental Exaltation' (273). Griesinger observed, having read the work of Baillarger and Falret, that mania and melancholia could alternate repeatedly, and might even follow the seasons (winter and autumn for melancholia), as he himself had seen in his own patients. According to Griesinger, the manic person gains 'relief' by demonstrating emotions through actions, and the 'affective sphere of the mind and the will' becomes more free. The manic, unlike the melancholic, is always expressing him or herself through the body, 'perpetually speaking, shouting, weeping, dancing, leaping storming etc.' (274). For Griesinger, however, the concept of unitary psychosis (all mental disease being along a single continuum) meant that he differed from Baillarger and Falret.

In this Griesinger echoed the pronouncements of the great Boerhaave from the previous century and Guislain in his own. For the French, circular insanity was a separate disease. Later in the century different psychiatrists would follow the French view of 'circular insanity' (the English phrase for disorders alternating between mania and melancholia) or Griesinger's notion of a continuum of severity.

As with other German physicians, Griesinger recommended individual treatment that included psychological and physical methods, even if the cause was solely organic. Early intervention and removal from the depressing environment (including occupation) were preferable. Like Pinel in France, Griesinger was active in asylum reform and argued for the integration of the mentally ill into society. The usual regulation of the non-naturals continued an ancient tradition, now reformed in terms of the nerves, and opium, purgatives, and sedatives had their place, alongside a humane moral management of the melancholic (although sometimes a stern manner might be better than constant consolation). Appropriate distractions from the obsessions or delusions of the patient and plenty of exercise balanced with adequate rest (and at least removal from previous causes of mental exhaustion) were all advised by Griesinger. As with other countries, access to certain remedies, such as hydrotherapy at fashionable spas and certain types of drug, was limited by class position. Clearly the melancholic rich could afford to remove themselves from their depressing social situation in ways unavailable to the lower orders.

British alienists held their own line on melancholia, although they were influenced by European and American views. James Prichard (1835), Bristol Quaker physician and later Commissioner in Lunacy, regarded melancholia as a form of

'moral insanity' (emotional disorder) and not one of the intellect. 'Morbid sorrow and melancholy' 'does not destroy the
understanding'.²² Mania, as in the eighteenth century, he saw
as 'raving madness'. Quakers had founded the humane York
Retreat, and the descendant of the first founders, D. Hack Tuke,
and John Bucknill's *A Manual of Psychological Medicine* (1858), the
psychiatric bible for many years, did not classify melancholia as
a form of 'emotional insanity' unless it was 'melancholia without delusions'.²³ Tuke agreed broadly with Esquirol that melancholia was 'a cerebral malady . . . without fever, and sustained by
a passion of a sad, debilitating, or oppressive character'. Unlike
Esquirol, Tuke did not regard delusion ('disorder of the intellect') to be an essential aspect of the condition (158). There were
six forms of melancholia: simple (no delusions); complicated
(with delirium or psychotic); acute, chronic, remittent, and
intermittent. As well as the usual psychological symptoms, he
cited a number of physical symptoms, including problems with
'uterine functions' in women and loss of libido in men (157–8).
Tuke also mentioned a form of depression that involved such a
'decided inaction of the intellectual faculties' that, rather than
being a simple depression, had gained the name of *melancholia
attonita*—a catatonic state. Interestingly, Bucknill and Tuke did
not attempt to organise their categories according to a higher
logic in the contemporary fashion, but attended to the listing of
conditions (Berrios, 395).

Bucknill raised a crucial point that is a recurring theme in
the history of depression: he argued that 'uncomplicated melancholia' is not different in kind from the normal grief and sorrow that are part of human life—merely in degree. However, he
did not side with other alienists who asserted that 'melancholia
is frequently a mere growth from a state of normal grief and

low spirits'. For Bucknill, some people were predisposed to the excessive reaction that would constitute melancholia proper—hereditary problems were the main factor, set off by a range of causes including 'all the moral causes of mental disease' like grief, loss, and anxiety. Predictably, the 'grand climacteric [menopause] of women' was also frequently accompanied by melancholy (309).

The causes of insanity in general were mysterious: not enough was known, said Bucknill, of the way the nerves worked and what laws might govern 'nerve-force'. He held that 'mental health is dependent upon the due nutrition, stimulation, and repose of the brain; that is, upon the conditions of the exhaustion and reparation of its nerve substance being maintained in a healthy and regular state' (342). He speculated that insanity could have physical causes in either an over-supply of blood and nutrition to the brain, or conversely an under-supply, over-exciting or under-exciting the cerebral functions in a manner reminiscent of Cullen and Brown's ideas of sthenic and asthenic excitability. Bucknill implied that melancholia would belong to the atrophied and drained brains that had been starved of blood. Bucknill's suggestions for treatment were of the standard exercise and fresh air type, with good food and tonic medicines, but did point out the need for sedative opiate therapy in the case of severe melancholia, when moral management might be ineffective. Bucknill was strongly against the use of restraints, and challenged American asylum superintendents to free their patients: his advice largely fell on deaf ears.

Later in the century, the pre-eminent British psychiatrist of his day and founder of the famous Maudsley Hospital in London, Henry Maudsley (1835–1918), roughly followed the native tradition in that he made it equivalent to Bucknill and

Tuke's 'simple melancholia' (without delusions).[24] This was 'insanity with depression', whereas a second form was 'melancholia with delusions' (today 'psychotic'). Out of the sad feeling of the second type of 'melancholic gloom' depressing ideas 'emerge dimly and shape themselves by degrees as positive delusions of thought' (188). These might include hypochondriacal thoughts as well as melancholia with a stupor. He did stress that his categories were fluid, as one kind of mental illness could easily evolve into another type, so that a flexible understanding of these different conditions should be adopted. *Simple* melancholia, in which the sufferer feels 'strangely and unnaturally changed…strangely isolated, and cannot take any interest in his affairs…is profoundly miserable and shuns society', could be associated with an idea which then leads to a delusion, thus becoming *partial ideational* melancholia. At the base of melancholia, said Maudsley, was a 'vast and formless feeling of profound misery'—rather than the delusion, which would follow after the crippling emotion (374).

As with other interpretations of melancholy, a long list of possible physical symptoms was provided by Maudsley: problems with the digestive system including loss of appetite; bad circulation and menstrual difficulties in women, whether suppressed or erratic; a slowing of the body's movements and sleeplessness. Maudsley had been influenced by Griesinger's statement that the diseases of the brain caused mental illness in general—so much so that Maudsley went beyond Griesinger in his insistence on this fact. Hereditary factors in the creation of melancholia were most important, with the theory of degeneration a significant presence here. As with Bucknill and Tuke, he considered the supply of blood to the brain—in both quality and quantity—to be crucial. Substances likely to enter the bloodstream, including alcohol and opium, were possible precipitating causes for melancholia. Problems in other parts of the body could affect the brain too, and the familiar role of powerful emotions or overwork might also provoke illness in that organ. Early intervention and removal of the patient from normal surroundings were thought to be helpful, as were moral treatment and baths, healthy diet, and judiciously dosed opium.

Notorious German psychiatrist and sexologist Richard Von Krafft-Ebing's *Text-Book of Insanity* (tr. 1904) was perhaps the most popular work on the subject in the final twenty years of the nineteenth century. Although known more widely for his case studies of sexual 'perversions', he also worked on less controversial matters. Melancholia was a 'psychoneurosis', or a disease state 'of the normal and robust brain'. He regarded melancholia as potentially reversible, having been caused by 'acquired diseases in individuals whose cerebral functions were previously normal'.[25] Because it was so difficult to identify particular causes in different cases, he argued that classification had to be achieved through grouping symptoms and the progression of the psychoneuroses, a method that has had significant influence on psychiatric method and practice to the present day. As to melancholia itself, its main element was 'painful emotional depression, which has no external, or an insufficient external, cause, and general inhibition of the mental activities, which may be entirely arrested' (286).

Krafft-Ebing followed previous thinkers in finding interference with the supply of nutrition via the blood to the brain to be a cause of a depressed and painful emotional state and 'inhibition of psychic activities, feelings, intellect, will' (286). The changes in the 'organ of consciousness itself' rather than merely a nerve that feels pain would have various psychological effects, he argued: 'the external world seems sombre and changed—in other colors'; the intellect was now slave to the emotions and became 'unable to retain in consciousness any other than painful and depressed images and ideas. The immediate result of this is monotony of thought and consequent weariness' (49). Now that the conscious process of thought was affected, the will was inhibited and the patient further depressed by the overwhelming disruption to the 'psychic mechanism' (50). Other physical effects, including sensory and muscular, followed from the lack of nutrition, and could result in hypochondriac tendencies, although the depression in itself was 'objectless' (51). Melancholia could be with or without delusions, and the content of delusions infinitely varied according to the social situation of the individual.

The most common form of mental illness was 'simple melancholia', according to Krafft-Ebing; 'it presents clinically great variety in the grouping of symptoms and the intensity of the disease'. Melancholia without delusions was the mildest form, and usually seen in private practice rather than in mental hospitals. It was quite often mistaken for other conditions like 'anemia, chlorosis, hysteria, neurasthenia, etc.' (293). From this point melancholia would increase in severity until it became delusional, and he noted religious and hypochondriac melancholia as 'especially striking and frequently observed' (301–5). This spectrum echoed Griesinger's idea of unitary psychosis.

Krafft-Ebing stated one of the cardinal concerns of the Victorian age when defining melancholia: 'the fundamental character of melancholia is that of absence of energy: passiveness' (289). In an age so concerned with the production of energy, there was a strong focus on the sources and drains on human energy. To counter this lack of energy, the patient should have 'complete' rest, physical and mental, should be protected from her or himself by surveillance; proper diet; relief from sleeplessness via opium; and symptomatic remedies like baths and opium—again.

Leading turn-of-the-century psychiatrist Charles Mercier (1852–1918) summarised the British nineteenth-century medical consensus on melancholia in Tuke's *A Dictionary of Psychological Medicine* (1892). Psychiatry had followed neurology's lead in claiming mental illness to be a disorder of the brain, the material body. Like authors before him, he emphasised the excessive nature of the sadness proportional to the circumstances that evinced it, and identified it as the primary symptom, along with a slowing of the physical constitution allied to 'defects of nutrition' ('a slackening, weakening, diminution of activity in the process of nutrition')—there might also be delusions.[26] He considered the prognosis for melancholia to be better than other conditions, and that its progress would be slow and usually starting in an otherwise healthy person. Only very gradually would dullness and lethargy progress into a full-blown 'morbid depression' (ii:788).

A simple melancholia might then develop into mania, and the idea that melancholia was the first state in all mental illness was upheld by many physicians. Mercier was keen to simplify the sometimes dizzying number of subcategories of melancholia, and stayed with the divisions of simple melancholia, melancholia with delusion, acute and chronic melancholia, active melancholia (accompanied by loud crying and physical movements), passive melancholia ('listless, lethargic and languid'), suicidal melancholia, and intervals of melancholy in other conditions (ii:788–9).

Mercier thought that melancholia was created through 'an inefficiency or slackening in the mode of working of the nerve-elements' (ii:791). 'Nerve energy' needed to be suitably active within the nerve tissue to produce a 'feeling of well-being'. If nerve energy or force was not sufficiently active, the general ability of the body to function was affected, and 'the defect of conduct, the passivity, the indolence, the lethargy of melancholia are dependent upon precisely the same alteration of nerve action as the constipation, the loaded urine, the foul tongue and the other physical symptoms' (ii:791). Such problems could come about though the predictable hereditary factor, and then various precipitating circumstances including physical maladies or changes (such as childbirth), 'exhausting exertion', and any number of causes of grief or sorrow, such as the loss of 'friends' and 'fortune' (ii:792).

Cure meant returning the nerve force to its proper level: diet and exercise helped nutritive absorption, thought Mercier, and—alarmingly to the modern reader—drugs such as 'iron, quinine, arsenic and strychnine', which supposedly invigorated the digestive system (ii:794). To combat insomnia, if the diet and exercise had not removed it, he thought supper with alcohol, or morphia or choral if necessary, would be effective. Vigilance against suicide in certain cases was, as usual, a given.

Such, then, was the state of play regarding melancholia, but another contender for the possession of depressive states came on the scene towards the end of this period: neurasthenia.

Neurasthenia

Neurasthenia now seems a quaint term—at least in the West—an influential oddity from the late nineteenth and early twentieth centuries. It is important for the story of depression because it could encompass many cases that would have been labelled as simple melancholia, and some which would be nearer to states combining melancholia and hypochondriasis. In its own time it was defined by the key American neurologist George M. Beard (1839–83) as 'exhaustion of the nervous system', a lack of 'nervous force'.[27] It provided a convenient label for the rag-bag of ideas floating around the medical ether concerning nervous exhaustion and depression. Over-excitation of nerve force might result in mania, epilepsy, or migraine.

If patients betrayed no sign of 'organic disease' then the nervous system might be suspected. The apparently endless list of symptoms might include: 'a general malaise, debility of all the functions, poor appetite, abiding weakness in the back and spine, fugitive neuralgic pains, hysteria, insomnia, hypchondriases, disinclination for consecutive mental labor, severe and weakening attacks of sick headache and other analogous symptoms'. This harks back to the diseases of the nerves so prevalent in the previous century, and inherits the Brunonian concept of asthenic excitability from the Romantic period—something one would have thought might become a dead letter in the later nineteenth century, but which chimed well with the industrial age's obsession with energy and its potential for exhaustion—even in the human body and mind. Beard regarded neurasthenia as a result of the fast pace of entrepreneurial urban life—particularly American—and as a uniquely nineteenth-century phenomenon. E. H. Van Deusen, who also coined the term independently in 1869, interpreted it as a disease of rural isolation as befitted his own situation in Kalamazoo, Michigan. Clearly this catchily named new condition was nothing if not flexible, and was used by different nations in different ways. The British, for example, were unconvinced that it was a new disease indicative of superior American civilisation, and saw the continuities with the English Malady of the previous century. Beard's American version excluded those unrefined by the height of (Anglo-Saxon) civilisation: blacks, native Americans, the lower classes, Catholics, most immigrants, and so on.

Clearly neurasthenia and melancholia were thinly divided, and in some cases not at all. Like melancholia, neurasthenia was often seen as the first step on the road to more severe mental problems, and, as with the logic of Krafft-Ebing on melancholy as an absence of energy, neurasthenia was evidence of exhaustion. In *Medicine and the Mind* (tr. 1900) Maurice de Fleury (1860–1931) explained melancholy and neurasthenia more or less interchangeably as affective or emotional conditions related to physiological states of over- or under-excitability: depletion of energy or nerve force versus an unhealthy lack of expenditure (in fact 'nervous bankruptcy' was a term commonly used by Beard and his followers).[28] Neurasthenia could also conveniently encompass those lower-level melancholic states that might not have come under the radar of institutional psychiatry, particularly those admitted to hospitals (see Fig. 13). There was a burgeoning market in home treatment—sometimes for good reasons connected with social status—for which a diagnosis of neurasthenia might prove very convenient. We have seen the case of Virginia Woolf already, in which 'neurasthenia' would save her from the asylum.[29]

De Fleury explained the mechanism by which the neurasthenic operated: if one hears news of the death of a loved one, 'by our eyes or ears, by the optic or auditory nerve, [it] project[s] strong vibrations to our nerve centres; and these vibrations themselves awaken and rudely destroy notions so firmly fixed…that the brain is overwhelmed by them and overwrought. Its vitality becomes exhausted and its tonicity is

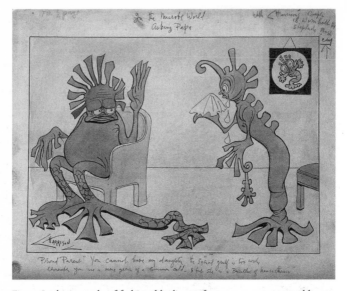

Fig. 13 In this comedy of fashionable disease from 1913 a common cold germ asks the father of a neurasthenia bacillus if he can marry her, but the class gulf between them is too wide. Neurasthenia allowed, as does depression, a range of mental states, from the voguishly sophisticated sensitivity of a Marcel Proust to the seriously debilitating misery of its full onslaught. (*Wellcome Library, London*)

lessened. Thenceforward the circulation grows languid, respiration becomes weak, the muscles are relaxed and work feebly, and the nerves of sensibility carry to the brain from the whole body the continuous idea of weakness, failure, powerlessness; the mind becomes conscious of this, with a vague and confused consciousness, and that is called grief' (263–64). Grief would become habitual, and melancholy would ensue. Hence 'melancholy is only a symptom of the disease of the vitality, an impoverishing of the circulation and a slackening of nutrition'. Again, the cure for such depletion of vital nerve force was the usual tonics to restore a healthy level of energy, and then to place the patient in 'useful and regular work' proportionate to the ability of the individual concerned (281). In contrast, the American 'cure' for neurasthenia, so notoriously satirised in Charlotte Perkin Gilman's *The Yellow Wallpaper*, had the effect of boring at least some women into lunacy, although it is impossible to know how many were cured by it. Other literary representations of neurasthenia are to be found in many texts, including Kate Chopin's *The Awakening* (1899), Frank Norris's *The Pit* (1903), Edith Wharton's *House of Mirth* (1905), Jack London's *Martin Eden* (1909), and Theodore Dreiser's *The 'Genius'* (1915). The British generally thought the American approach too extreme and tended to recommend rest in weeks rather than months. De Fleury himself rejoiced in the opinion that 'melancholy depression' could be cured effectively, although here he may have been referring to lower-level depression as we would now understand it.

The British accepted neurasthenia as a disease in the 1880s, and with it revived the notion of depressive genius. As Thomas Savill put it: 'many neurasthenic subjects are persons of very considerable intelligence and brilliancy, who, therefore, take a leading part in society'.[30] John Addington Symonds thought his literary creativity due in part to his 'high degree of nervous sensibility'.[31] The logic of neurotic genius, in the eighteenth-century sense, was given new validity at the end of the nineteenth century, despite the fact that it had to contend with the muscular Christianity and theory of degeneration of the second part of the century.

One major difference from the previous century's interpretation of melancholic inspiration was the recognition, by the end of the century, that the lower classes might also suffer from nervous diseases. Even the great psychiatrist and neurologist

Charcot in France had pointed out the presence of working-class patients. Nevertheless, the social orders were treated differently: whereas physical exhaustion might be blamed as the cause of a lower-class person's neurasthenia, moral agents—grief, anxiety, intellectual overwork—tended to be assumed as the drivers for middle- and upper-class diseases. Writers in Europe and America depicted neurasthenics from all classes, from Edith Wharton's 'old money' Mrs Peniston in *The House of Mirth* to the impoverished Mrs Frome in *Ethan Frome* (1911).

Neurasthenia continued in its catch-all popularity until after the First World War (although as *shinkeisuijaku* it was to make its way to Japan and flourish there to the present day, in that culture it lacks the stigma of a psychiatric diagnosis). The advent of Freud and his ideas about neurosis stemming from unconscious sexual conflicts made the previously satisfying somatic basis of neurasthenia untenable. Now it was a potential embarrassment to admit that one's emotional state might have sexual drives at its root.

At the end of the nineteenth century, then, the concept of melancholia and the less technical 'melancholy' was clinging on to its place in the popular as well as psychiatric consciousness, but the term 'depression' was moving more rapidly towards its twentieth-century meaning. The transition from melancholy to depression was by no means smooth and, as we have seen, other names—such as neurasthenia—might obtrude to describe depressive states, but we arrive in the modernist period with a sense that the death of the Victorian period was to mean the death of melancholy. In the next chapter we will see how the 'new' depression came into full existence, and how it was consolidated and maintained up until the end of the new century.

V

MODERNISM, MELANCHOLIA, AND DEPRESSION

Nothing is funnier than unhappiness, I grant you that.
(Nell, in Samuel Beckett's *Endgame*, 1957)

B eckett is known as the last of the Modernists, those who broke away from the certainties and constraints of the immediate past. His attitude to depression also reflects the loss of God in twentieth-century Western culture. Beckett, who suffered from depression himself, saw life as absurd and a condition in which suffering is inevitable— 'You're on earth. There's no cure for that' (*Endgame*). This gloomy perspective differs from the late part of the twentieth century, however, in that we now expect to avoid misery and illness, physical or mental. Beckett's world view is much more medieval, but without the consolations of religion. Taken out of context, Nell's words are an oxymoron, but in the absurdist universe of the twentieth century, melancholia and depression acquire new forms, both cultural and medical. Beckett's despair is in part existential, but it is also funny precisely because of the absurdity of life in a century that witnessed two world wars,

Hiroshima, and the Holocaust. Beckett is indicative of broad cultural trends, but the definition and treatment of depression have been surprisingly varied and contested in the modern world, beginning with the split between psychoanalysis and mainstream psychiatry.

The Victorian psychiatry of the nineteenth century shifted into the modernism of the early twentieth via the evolution of the thinking of two major influences on the practice of the twentieth and twenty-first centuries: Emil Kraepelin and Sigmund Freud. Freud's psychoanalytical concept of depression has maintained its influence in subsequent artistic and literary representations, but Emil Kraepelin's (1856–1926) more 'scientific', biomedical approach to depression was to become much more significant in the development of psychiatric practice, certainly as it feeds into late twentieth-century definitions of depression in the *Diagnostic and Statistical Manual of Mental Disorders* (DSM) and the *International Classification of Diseases* (ICD). He has been called the 'father of modern psychiatry', despite being little known by the general public.

Kraepelin: towards modern psychiatry

Kraepelin's overarching category of manic-depressive insanity swallowed up the majority of what had been previously thought of as melancholias, and has also had a continuing influence up to our own time. A German psychiatrist, professor and head of department at the University of Heidelberg, he had been influenced by Griesinger's idea that diseases of the mind stemmed from diseases of a physical organ, the brain, and put his emphasis on that biomedical framework that has become so significant recently. Kraepelin, like neurologists of the nineteenth century, hoped that future clinical explorations would reveal the true physical causes of mental disease via brain lesions. Freudians too claimed scientific validity, but Kraepelin has come to be associated with the neurological and biomedical approach to depression.

Kraepelin invented categories of mental illness from groups of symptoms and the progression of disorders. His big division was between those conditions that might deteriorate and require continued treatment in an asylum, such as *dementia praecox* (schizophrenia to us), and those conditions that were usually episodic and could remit, thus allowing return to the home and general society, such as manic depression (nowadays bipolar disorder). These non-deteriorating disorders were affective, of the emotions, whereas the deteriorating ones were disorders of thought. His observations of a large inpatient population and meticulous case records helped him in his influential classification of mental disorders.

Kraepelin's thinking on melancholia and depression began in the last decades of the nineteenth century and evolved throughout that time into the theory that was to be so significant for the twentieth. For Kraepelin, mania and melancholia, or now the depressive state, were part of a single continuum of mental illness: '*Manic-depressive insanity* takes its course in single attacks, which either present the signs of a so-called manic excitement (flight of ideas, exaltation, and overactivity), or those of a peculiar psychic depression with psychomotor inhibition, or a mixture of the two states.'[1] Here Kraepelin contrasted triads of symptoms to show the fundamentals of the respective conditions: mania with flight of ideas, exaltation, overactivity, and depression with inhibition of thought and physical processes, and depression of feelings.

The *depressive states*, as Kraepelin put it, manifested themselves in terms of increasing severity—again, a familiar idea from the previous century, but influential in the twentieth. First came *simple retardation*, the 'mildest' form, minus delusions and hallucinations and characterised by a gradually increasing sluggishness and withdrawal of interest in the external environment. Memory started to fail and patients complained of feeling 'tired and exhausted'. Their minds remained clear, however, despite this mental 'retardation'.[2] That said, this is a gritty picture of depression that does not fit with the previous idea of melancholic inspiration. The 'emotional attitude' is one of 'uniform depression', in which 'the patient sees only the dark side of life'. These poor souls were failures in professional life, had lost faith in religion, threatened suicide but rarely seriously attempted it. They were despairing and had no 'fit' with their general social environment. This stage could last for a few months to more than a year.

In his *Textbook of Psychiatry* (1909–15, 8th edition) Kraepelin invoked the literary source of Goethe's Romantic character of young Werther to describe the detachment of the depressive from the world, or 'depersonalisation': 'I stand as though in front of a cabinet of curiosities, and I see little men and little horses moving about in front of me, and I often ask myself whether it is not an optical delusion.'[3] This was part of the first category of the depressive states, labelled by Kraepelin as *melancholia simplex*. As ever, creative literature provided a ready means of expressing the nuance of psychiatric observation, albeit framed by generic constraints.

The next stage in depression, *melancholia gravis*, added hallucinations and paranoia or self-loathing to the previous symptoms, with hypochondriacal delusions also possible. Again, there was clarity of consciousness, but 'a constant tendency to revert to their depressive delusions'.[4] 'The patients see figures, spirits, the corpses of their relatives; something is falsely represented to them, "all sorts of devil's work". Green rags fall from the walls; a coloured spot on the wall is a snapping mouth, which bites the heads off children; everything looks black.'[5] 'Ideas of sin', overwhelming guilt, might plague the sufferer, with religion being 'a peculiarly favourable soil for self-accusation.'[6] 'Ideas of persecution' were closely related to notions of transgression—the feeling that the patient should be punished and that she or he was universally condemned. Following from this, 'Paranoid melancholia' could develop, 'when ideas of persecution and hallucinations of hearing are frequently present and sense remains preserved'. Yet another subset was 'Fantastic melancholia', a state characterised by 'abundant hallucinations', varying greatly but keyed, as delusions usually are, to the culture of the time: 'The patient is electrified by the telephone, is illuminated at night by Röntgen-rays, pulled along by his hair; someone is lying in his bed; his food tastes of soapy water or excrement, of corpses or mildew'.[7]

The physical symptoms included the literal feeling of de-pression—'a feeling as if there were weights on the chest'—as well as the common features of the melancholic like constipation, poor appetite, and disrupted sleep, racing heart and 'sallow' skin, 'lustreless eyes'—all indications of the drain on body as well of mind of a disease originating in organic malfunction. This type might come on suddenly after illness, low spirits, or even 'a short period of exhilaration'. It might last from six to eighteen months, with some 'partial remissions' and 'very gradual improvement'. In the round, manic-depressive states boded badly for the lifetime of the individual in terms of recurrence, but had a positive outlook for the remission of the particular attacks.

The final subset of 'Delirious melancholia' was equated by Kraepelin with the familiar category of 'melancholia with delusions' from the previous century, as in Griesinger. This 'depressive insanity' totally disorientated the sufferer, with 'a profound visionary clouding of conscience'. The patients had 'numerous, terrifying hallucinations, changing variously, and confused delusions are developed'. Some visions were apocalyptic and religious, others physical (the patient changed sex, had cancer and ulcers). During these experiences the patients were profoundly disorientated: they 'know absolutely nothing any longer, give contradictory, incomprehensible, unconnected answers'.[8]

A third category of the depressive states was that of *stupor*, 'characterized by numerous incoherent and dreamlike delusions and hallucinations, with a pronounced clouding of consciousness. This form rarely appears alone, but usually forms an episode in the course of the other forms' (*Clinical Psychiatry*, 303). Kraepelin also introduced the category of 'involutional melancholia', a depressive disorder thought to strike in late middle age or old age and to be accompanied by paranoia. He later added this separate category to the manic-depressive disorders as a mixed state, now considering them to be better in prognosis than he had previously thought.

It is no surprise to find that Kraepelin echoed the previous century in identifying the primary cause of these mood disorders to be hereditary predisposition—in up to eighty per cent of cases. Women were apparently more susceptible than men, a theme to which we will return. Kraepelin placed a heavy emphasis on internal factors rather than external events like the death of a loved one, romantic issues, or money problems. Even these events, which might seem on the surface to have caused depression, would in fact be due to innate internal factors.

Kraepelin is important for the progress of twentieth-century psychiatry because of his attitude that one could infer the existence of a mental disorder and its underlying physical cause by carefully following symptoms over the course of an illness. His version of depressive symptoms has constituted the basis for diagnosis and classification for depressive disorders up to the present. One problem that Kraepelin's theory stored up for later decades was the way in which his idea that the dizzying array of symptoms displayed in manic-depressive illness was the manifestation of one underlying illness condition, and that some symptoms might be surface manifestations of that deeper condition. Thus, he decided that some people who were only showing signs of depression had manic depression—despite there being no evidence of mania. Similarly, his assumption that over time patients with mood disorder might display depressive, manic, or mixed states led to this tendency to assume a deep pathology to symptoms that might not necessarily be interpreted as being serious. Even mild symptoms that could be natural sadness due to external events could be seen by Kraepelin as the precursor to more severe illness.

In terms of therapeutics, Kraepelin was not particularly innovative: he recommended the rest cure outside hospital as a basis for treatment in the milder cases, and, in severer cases, committal to an asylum. He had been more often involved with these severer kinds of cases in his career as a physician, having started in a Munich asylum before pursuing his interest in psychotic disorders in the Psychiatric Clinic in Munich. Change of psychological and physical environment was important, whether it be a placement in a cheerful family or in a community. Those providing care (which should be constant) needed to be gentle and reassuring. Stopping visits from relatives was also a good idea, in order to prevent emotional stress on the patient. Building up the constitution through bed rest and a nutritious diet was necessary, and warm baths—which he thought preferable to sedatives—useful for insomnia.

As ever, potential suicide required vigilance, and the asylum was the best place to ensure this. In therapeutics, then, Kraepelin championed the asylum, but his treatments were not out of the ordinary for his time, and, it might be argued, his concept of manic-depressive disorder was well prepared in various contexts in the nineteenth century before he pulled the different strands together. Nevertheless, in his method of classifying syndromes (usual patterns of symptoms) and placing mental illnesses in clusters based on those syndromes, he set the template for modern psychiatry, and showed the failure of the old classificatory methods, which merely involved comparing major symptoms.

Freudian alternatives

Sigmund Freud's (1856–1939) essay Mourning and Melancholia (1917) defined melancholia as a response to loss in childhood, often involving repressed anger directed against the self. Depression was not yet 'depression' for Freud: he used it as a descriptive term rather than disease category. For Freud, melancholia was a feature of unconscious conflict: an idea that has had an extraordinary afterlife. Like Kraepelin, Freud began his work in the nineteenth century, but his influence in the twentieth century has been monumental. For Freud, mental illness came from mental causes, not necessarily physical ones. Rather than blaming depression on one's mother's defective nervous constitution, or some physical injury, Freud focused on the operations of unconscious drives, hidden desires, and conflicts in the psyche. Freud too was concerned with energy, but he considered that the energy of the libido and other aspects of the mind could be converted into anxiety, depression, or other physical consequences that followed on from the psychological conflict. Although the idea of the passions as one of the 'non-naturals' being able to cause melancholia had been in existence since the Ancients, Freud broke from previous explanations of the emotions having an effect on the body and mind by introducing the concept of the unconscious, a part of the mind profoundly unknown to the conscious mind (so *un-* not *sub-* conscious). The process of repression, said Freud, prevents certain psychological conflicts from being present to the conscious mind, but that very process of repression could cause those conflicts to fester in the unconscious and eventually manifest themselves in mental and physical illness.

Freud's follower and collaborator Karl Abraham (1877–1925) aided Freud's thinking on depression by writing the first consideration of depression from a psychoanalytic perspective. A German psychoanalyst, he had learned of Freud through his position at the Burghölzli Mental Hospital in Switzerland, where Carl Jung had also worked. He was able to provide detailed clinical data on depressive conditions, and observed that the feeling of depression was as commonly distributed among the neuroses and psychoses as anxiety, and that both anxiety and depression were caused by repression of unconscious conflicts, whereas fear and sadness were normal emotions with clear causes. In his 'Notes' of 1911 Abraham saw depression as a product of anger directed against the self rather than being expressed outwardly against another person.[9] By contrast, normal mourning over a 'lost' person was the result of a *conscious* obsession with the lost person. Release of the depressed person's anger became one of the goals for treatment in the psychoanalytic tradition. Later, Abraham was to be the analyst and teacher of other significant figures in the Freudian line, such as Sándor Radó and Melanie Klein.

Freud expanded Abraham's initial ideas on depression in his seminal essay 'Mourning and Melancholia'.[10] Mourning was normal grief, a natural process in reaction to the loss of a loved object, and not requiring medical intervention. In this sense, mourning was depression with a cause. All that was required for the mourning person to recover was to allow the grieving process to take place without any outside interference, even from a psychoanalyst. Melancholia, on the other hand, was depression without cause, even though it had similar symptoms to mourning. Both conditions relied on different psychodynamics, one conscious and the other unconscious. The melancholic was unaware that she or he was transforming anger at an earlier loved object (such as a parent, with whom the melancholic strongly identifies) into self-loathing, with the resultant symptoms of sadness, lack of pleasure and energy, and withdrawal from the external world. Both conditions were focused on loss, according to Freud, but the nature of that loss was very different.

Freud speculated that the libidinal energy of the infant, which at first was directed to the ego, became heavily identified with the loved 'other', to the extent where the ego felt that it had incorporated the other into itself. When adult woe occurred, said Freud, the infant experience returned, with the ego attacking the introjected 'other' or 'object'. The consequence of this self-directed accusation meant that melancholia would ensue. Freud argued that the odd apathy of the depressive response proved that the love object—now fantasised as part of the ego—was the actual focus of the redirected anger. The melancholic's unconscious loss—the feeling of an empty ego—could only be remedied by allowing that unconscious anger to become conscious and properly directed towards its real object.

Freud's new emphasis on psychic loss rather than physical imbalance or malfunction is accompanied in the essay by some echoes of older traditions, including the idea that melancholics might have greater abilities and insights than those of others. Freud picks Hamlet as an exemplary melancholic and states that the depressive person 'has a keener eye for the truth than other people who are not melancholic' (246). The discourse of sadness without cause is also repeated via Freud's admission that 'Melancholia, whose definition fluctuates even in descriptive psychiatry, takes on various forms the grouping together of which into a single unity does not seem to be established with certainty' (243). This also is a fecund passage for those of a postmodernist persuasion who have used Freud as a way of demonstrating the radical instability of both language and reality, as we will see in the final chapter.

Freud's legacy

The years after the Second World War to the 1970s marked the age of psychoanalysis in America, inspired in large part by the influx of European refugees from the war and the groundwork laid by the writings of Freud and his acolytes before that period. Sándor Radó (1890–1972), a Hungarian psychoanalyst, was one of Abraham's students and had met Freud before the First World War. Radó helped develop this European tradition by going on to analyse influential thinkers like Otto Fenichel and Wilhelm Riech. Later in his career he fell out with Freud in a power struggle over the future of American psychoanalysis, a new terrain that Freudians hoped to exploit via the New York Psychoanalytic Society, which established its first Training Institute in 1931, with Radó as director. Radó considered that melancholia was 'a great despairing cry for love'. He introduced a concept that echoes in popular psychology down to the present day: that 'the most striking feature' of depression was 'the fall in self-esteem and self-satisfaction'.[11] In his early thinking Radó distinguished between *depressive neurosis*—a less severe form in which the patient is conscious of low self-esteem and attempts to conceal the fact, and *melancholia*, in which there are loud 'delusional self-accusations' and it approaches a psychotic depression (i:48).

Radó's rather uncomplimentary view of depressives placed narcissism at the heart of the matter: like a small child, they seek approval and love from their various love objects and become angry when that love is not forthcoming. They then go through a pattern of 'guilt, atonement and forgiveness' within their own mental life, as a conflict between the ego (self) and superego (conscience or conscious moral self). This strangely Catholic framework stems from the 'rage, hunger, drinking at the mother's breast' triad from infancy (53). Such a hunger tokens the loss of love that in later life prompts the depressive state, and in more extreme cases of true melancholia threatens to move from *neurosis*, in which there is still contact with reality and the object, to *psychosis*, in which the bonds with reality are loosened or broken.

In the 1950s Radó changed his ideas on depression as he developed his 'adaptational psychodynamics', where he concluded that 'the entire depressive process must be evaluated from the adaptational point of view, and interpreted as a process of miscarried repair'. Guilty and submissive fear about regaining the mother's love would result in 'retarded depression'. If rage was the driver, then 'agitated depression' ensued (238).

Otto Fenichel (1898–1946), a Viennese psychoanalyst who met Freud in 1915 and then moved to Berlin under Radó's tutelage, usefully summarised developments in psychoanalysis up to 1945, and pointed out that depression was an aspect of most forms of neurosis, but in extremis 'it is the most terrible symptom in the tormenting state of melancholia'.[12] Like Radó, he saw dependent narcissism, loss, and low self-esteem as the fundamentals of depression. Low-level depression was a 'warning' that a severe depression might be coming if 'the vital supplies are lacking' for the 'orally dependent individual', or someone who had been denied food and oral stimulation at the 'oral stage' during the first year or two of life. The borderline in the continuum of depressive states between neurosis and psychosis ('where the conflict has become internalised') was not clearly divided, Fenichel also noted.

Melanie Klein (1882–1960) was trained in the Freudian tradition, having been born in Vienna, but ended her life living in London and strongly influencing the general psychoanalytical tradition through her concept of 'object relations'. Focusing on early-years child development (she was a mother of three and had analysed them herself), Klein theorised that the infant attached ambivalent feelings of both love and hate towards its first 'objects'—essentially the mother and her 'part-objects' such as the breasts. The child 'incorporates' aspects of the object into the self or ego, and the ambivalence of the child towards the other can play out in conflicts within the self, in a complex process of projection and 'introjection' of feelings. Klein's idea of the 'depressive position', first broached in 1935, suggested that the time of weaning involved intense feelings of separation, loss, and distress.[13] Such a stage was similar to Freudian melancholia, and crucial in the development of the child: events in adulthood might trigger those depressive feelings again.

To overcome the depressive position in adult life, the psychoanalyst needed to identify and cause the patient to 'introject' positive feelings from the 'good' object or (representation of the) mother, feelings of love and security. The focus on the mother–infant relationship was both a strength and a weakness of Klein's work, both highly influential yet also subject to criticism in attributing such a great importance to the possible feelings of the infant and the prominence of the role of the mother. It is not coincidental that Klein's own depression evidently stemmed from her difficult relationship with her own mother.

In mid-century, Edward Bibring (1895–1959), a Jewish psychoanalyst born in Galicia but trained in Vienna, pursued the idea that lost self-esteem was at the heart of depression, arguing that the 'simple' as well as 'melancholic' forms of depression were merely stages in the attempt to restore self-esteem though different psychic mechanisms.[14] Depression was 'the emotional expression (indication) of a state of helplessness and powerlessness of the ego, irrespective of what may have caused the breakdown of the mechanisms which established his self-esteem' (24). Depression resulted from the inability of the ego to live up to its aspirations to be worthy, strong, good, and loving, according to Bibring.

Less fixated on the role of oral gratification—or lack of it—stressed by previous thinkers, Bibring considered that an infant ego's early experience of its helplessness, or inability to provide 'supplies', was the most prominent predisposing factor for the triggering of depression in adult life, when 'the painful discovery of not being loved' could reactivate infantile feelings. This was the fundamental mechanism, with variations on depressive states as 'complications' of that basic process. Importantly, Bibring also commented on a feature of depression that has played out controversially in its history: the narcissistic secondary gain to be had from depression, such as increased attention or affection. Bibring emigrated to the United States (via Britain), as so many Jewish inhabitants of Austria had to do because of the Second World War.

Edith Jacobson (1897–1978), in a similar trajectory, was born in Germany, trained in physical rather than psychological medicine initially, and died in New York, her emigration prompted by her imprisonment by the Nazis for refusing to divulge information on a patient. Her paediatric internship at the University Hospital in Heidelberg led to her developing a special interest in child psychology and psychoanalysis. She was less happy to have neurotic and psychotic depressions on a continuum of severity, and considered psychotic depressions to be neurophysiological in origin, although layered over by psychological causal factors, whereas neurotic depression was not so. Manic depressions were psychotic at all times. She thought a major depression was generated by 'a lack of understanding and acceptance by the mother' which would then reduce the child's self-esteem, thus generating a conflict that turned aggression against the self and created a feeling of helplessness and hence depression.[15] The severity of the depression would be dictated by the 'intensity of the hostility' and the length of time of the frustration.

The diagnosis of depression in psychoanalysis tended, with variations, to focus on the role of women as mothers and early sexual development, and the role of men as patriarchs in an Oedipal scenario. Feminist critiques of this situation mushroomed in the 1970s and 1980s, but at that time the influence of psychoanalysis had waned and the biomedical model was taking over, with the emphasis on symptoms. Gender, however, would remain central to the diagnosis and treatment of depression.

Psychoanalysis and psychiatry: the American way

The first part of the century saw the rise of a blend of psychoanalytic and Kraepelinian approaches across the Western world, although it is largely within American and British cultures that the rest of this book will move. While the impact of Europe remains crucial, we will be tracing out the ultimately dominant effects of this research and practice on a global scale via the development of the American market.

In the United States before mid-century the 'biopsychosocial' approach of Adolf Meyer drove a more patient-focused method in mainstream psychiatric practice. Meyer worked with Kraepelinian notions in a pragmatic way while distancing himself from Freudian psychoanalysis (the psychoanalytic revolution in America would be largely after the Second World War). In a move away from the idea of discrete disease separate from a person's individual circumstances, Meyer argued in 1905 that the category of melancholia should be replaced by depression.

Meyer, a Swiss-born psychiatrist (1866–1950) and contemporary of Kraepelin and Freud, enters our story of depression as a man who led psychiatry in the United States for the first forty years of the twentieth century, and thus had a profound influence in the way it was understood and treated. He came to be increasingly critical of Kraepelin's way of defining depression and mania, and found it to be too inclusive. He pointed out that 'many *depressions*' did not belong to the groups within the manic-depressive rubric and suggested a number of such depressions that should be differentiated from Kraepelin's categories.[16]

He also complained about the nomenclature, stating in 1902 that 'in its current use melancholia applies to all abnormal conditions dominated by depression', and in 1905 suggested that he was 'desirous of eliminating the term melancholia, which implied a knowledge of something that we did not possess, and which had been employed in different specific ways by specific writers. If, instead of melancholia, we applied the term depression to the whole class, it would designate in an unassuming way exactly what was meant by the common use of the term melancholia; and nobody would doubt that for medical purposes the term would have to be amplified so as to denote the kind of depression...We might distinguish the pronounced types from the simple insufficiently differentiated depressions. Besides the manic-depressive depressions, the anxiety psychoses, the depressive deliria and depressive hallucinations, the depressive episodes of dementia praecox, the symptomatic depressions, non-differentiated depressions will occur' (568). This was one of the nails in the coffin for melancholia, as the baggage it carried was more trouble than the word was worth. It must be noted that it was not until the second half of the century that depression really achieved its modern meaning and usage. This shift also illustrated Meyer's pragmatic approach to treating mental illness: he was adaptable and willing to innovate or draw from different influences (both Kraepelin and Freud) in order to arrive at a better understanding of his subject.

Instead of Kraepelin's syndromes, Meyer suggested 'reaction types', based on the notion that mental illness is 'a faulty response or substitution of an insufficient or protective or evasive or mutilated attempt at adjustment'. Meyer had become frustrated with explanations forced to hypothesise hereditary predisposition, changes in cells, nerve weakness, and a host of other factors 'which we cannot reach or prove' (599). If a depressive reaction, part of what he called 'the simple depressions', 'more or less, excesses of normal depression', was regarded a reaction to a situation gone wrong, then one would be able to think about how one could deal with 'modifiable determining factors' that might bring cure or improvement. This way, the psychiatrist would be in 'a live field' rather than condemned to 'neurologizing tautology'. In short, Meyer's approach promised liberation from treatments and theories of the previous century.

The six disorders or reaction types Meyer suggested were: reactions of organic disorders; delirious states, essentially affective reactions; paranoic developments; substitutive disorders of the type of hysteria and psychasthenia (a condition now passé that is not unlike obsessive–compulsive disorder); and types of defect and deterioration. The 'simple depressions', as well as manic depression and anxiety, were to be found within the affective reaction types (600). Meyer had a penchant for coining new terms, some of which caught on, like psychobiology, and some that did not, like ergasiatry (for psychiatry) and ergasiology (for psychobiology), both of which were supposed to imply the functioning or 'work' of the individual. 'Psychobiology' was an approach that detailed the particular psychological, social, and biological features of a patient's case, rather than limiting analysis to one single factor. The idea of the mental functioning of the depressive being enmeshed with the working environment and process led Meyer to propose occupational therapy as one aspect of his therapeutics. For Meyer, one's everyday activities were a crucial factor in determining mental (and physical) health.

This adaptability was a strength of Meyer's approach: he argued for a meticulous history of the individual's personal situation and 'reaction-set'. Each person had a specific response to certain events and environments—if the causes for the reaction could be analysed, then a way of developing an appropriate adaptation rather than a maladjustment might be found. Because of his suspicion of previous unprovable theories, he treated patients pragmatically, rather than being led by one or other theory. Hereditariness bred pessimism in therapeutics: how could one escape from one's genetic destiny? Fortunately, Meyer broke from this attitude and proposed a therapy that would be 'a service in behalf of the patient' and which would use the 'assets' of the patient in a process of collaboration: 'The psychiatrist—the user of biography—must help the person himself transform the faulty and blundering attempt of nature to restore the balance, an attempt which has resulted merely in undermining the capacity for self-regulation.'[17] The problem with this approach was that it was so individualised that it threatened to prove everything and nothing: how could one make generalisations about disease if it was so heavily focused on a single patient?

Meyer's general therapeutic approach, as one can glean from his attitude to the patient, was humane and used psychotherapy as well as more physical interventions such as hydrotherapy, occupational therapy, and the usual attention to rest and nutrition. For the depressive patient, the physician needed to 'offer...a sense of security by communicating an understanding based on his personal knowledge of him and of the situation' and 'to avoid inducing any antagonistic attitude which would interfere with further unburdening'.[18]

Alongside the use of psychoanalytic concepts and therapies this period saw new somatic (or bodily) treatments being developed, some of which would end up being used on depressives: the history of drugs and electroconvulsive therapy (ECT) we will see later, but it suffices to say that by the 1940s some neurologists and psychiatrists were devoting more energy to physical treatments, including shock therapies: at this time both somatic and psychoanalytical approaches coexisted, although not always happily. One clinical psychiatrist writing in 1939 hoped that the success of insulin shock therapy and metrazol (a drug which in high doses induced convulsions) would get rid of Freudian psychoanalysis. Others complained that 'office' psychiatrists never dealt with the serious cases to be found in the asylum. In practice, both forms of therapy could be used at the same time.[19]

The influence of the world wars, particularly the Second World War in the American context, led to an expansion of the influence of psychiatry—especially in its psychoanalytic form—into the general population. Importantly for the later expansion of the diagnosis of depression, the apparent success of psychiatrists in treating soldiers suffering mental problems in the war (and returning them to the fighting) suggested that psychiatrists could help the general public. The key point here was that soldiers were not to be stigmatised as diseased, but merely reacting to combat fatigue just as one would with a physical injury. Treating such soldiers made psychiatrists seem like any other doctor: if they could treat soldiers, who were 'normal' people rather than extreme cases of mental illness needing to be placed in the asylum, then they could treat the general public, who might be subject to psychological problems now relieved of such stigma.

By the 1950s, more people were familiar with the work of psychiatrists and, thanks partly to popular magazines, had absorbed a popular psychological vocabulary that would serve them well as potential consumers of mental health services and products. This shift was helped by social psychiatrists who proclaimed their mission to bring better mental health to the public in the same way that the rest of the medical profession was able to do. This movement came to the fore after the Second World War and, in contrast to the focus on the individual promoted by psychoanalysis, examined the role of social factors in mental illness, the Midtown Manhattan Study (1954) being a landmark analysis. War psychiatry had illuminated the effect of external factors on the individual's mind, and the rise of sociology as a discipline dovetailed with this interest in social forces. A further effect of the Second World War was to promulgate the belief that all people were potentially at risk of mental illness as everyone potentially had a 'breaking point' triggered as much by environmental factors as by the individual's psychological makeup, and this encouraged the subsequent social psychiatry of the 1950s and 1960s, in which screening programmes that had been used (unsuccessfully) to detect unsuitable candidates for the army were now used on a general civilian population in the community surveys of the 1950s. Despite the fact that military screening programmes did not prevent psychiatric casualties or indeed include many potentially good soldiers, the general principle of a population 'at risk' of mental illness was strong enough to persist into civilian psychiatry after the war.

These various developments set the scene for the explosion of the mental health market in the second half of the twentieth century, although the early years would be more concerned with anxiety than depression.

VI

———∞∞∞———

THE NEW DEPRESSION

This chapter charts the rise of the 'New' Depression, a concept of pathological sadness that arose in the late 1970s and early 1980s and has continued to the present day. This concept of depression has been largely defined by a biological model most famously exemplified by its apparent amenability to treatment with Prozac, the magic bullet that came to prominence in the final decade of the twentieth century (see Fig. 14).

The American *Diagnostic and Statistical Manual of Mental Disorders III* (DSM-III) was published in 1980 and became the bible of the New Depression by providing a much wider clinical definition, based almost entirely on symptoms, of what it meant (and means) to be depressed. A combination of the politics of medical research and professional expediency led to the production of the DSM-III and, with it, a break with previous models of depression—a break that had real consequences for millions of sufferers.

The Freudian and Meyerian forms of psychiatry dominated ideas about depression until about 1970, leaving Kraepelin

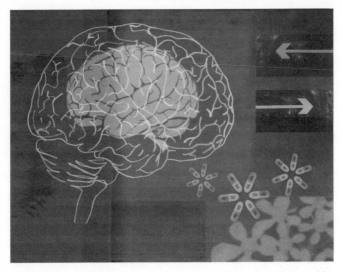

Fig. 14 'The brain and Prozac' by Rowena Dugdale—artwork indicative of the biomedical view of depression. Prozac is said to work by increasing the level of serotonin in the synapse (a bridge from one nerve cell to another for electrical or chemical signals), but there is much dispute about this process, including whether it works at all. (*Rowena Dugdale, Wellcome Library, London*)

largely out in the cold. Sharply defined disease categories were not necessary because the emphasis was on the way in which the conflicts of the unconscious mind could be resolved, rather than the protean symptoms to which they gave rise. Most patients in the United States paid for their own treatment beyond the asylum, so there was little involvement of third parties (such as insurance companies) demanding specific diagnoses of certain diseases. The situation changed after 1970, however, as the psychiatric profession came under pressure to provide a greater degree of certainty (or apparent certainty) than had hitherto been the case.

It's all in the symptoms: before the DSM-III

Before Kraepelin's re-emergence in DSM-III, some British researchers were developing his ideas through a statistical analysis of symptom clusters in an attempt to ascertain whether depression was one or many disorders. The use of statistics for medical purposes had been developing in the early twentieth century and mental health researchers had adapted statistical techniques to identify different groups of patients and diseases in institutional contexts at first, and after the world wars in community practice. Aubrey Lewis, a psychiatrist working with sixty-one patients at the London Maudsley Hospital in 1934, stated that the distinction between endogenous (of internal causation) and exogenous (externally motivated) depressions was illusory because most endogenous depressions did actually have an external precipitating factor and also that exogenous depressions were usually bound up with previous precipitating factors. Another small group of researchers claimed that they could not disentangle sufficiently distinct symptom patterns to show different types of depression. Both Lewis and these researchers showed evidence that seemed to confirm Kraepelin's argument that depression was one single disease and that the existing division between neurotic and psychotic was unfounded.

This was not a message that was accepted in the mainstream of psychiatric practice in either Britain or the United States, however, in which Freudian and Meyerian (context-based) ideas dominated from the 1920s to the 1970s. Psychotic depressions apparently responded to treatment in distinct ways: electroconvulsive therapy (ECT) and imipramine (the first 'tricyclic' antidepressant developed in the 1950s) had much clearer effects on the psychotic patient than the less severe 'reactive or neurotic' depressive. It was the case, however, that the term 'endogenous' shifted meaning because it was difficult to identify a depression of any sort that did not have some kind of contextual trigger. 'Endogenous' came to refer to a pattern of symptoms deemed to be more serious and/or psychotic than the milder neurotic type. The term 'neurotic' itself came to dominate 'reactive' for the same reason: that all kinds of depressions seemed have an environmental stress that predisposed the sufferer to the illness.

Although these researchers claimed to be emulating Krae-pelin's methodology, they actually tended to study clusters of symptoms in a single sitting, without the meticulous recording of the disease progress and contextual history that constituted Kraepelin's technique. 'Factor analysis' was a new way to exploit statistical data about symptoms of depression by analysing how far single symptoms clustered with other symptoms, and what got lost in the brave new world of statistics was the subtlety of Kraepelin's chronological approach. Only symptom patterns mattered now, while important matters like the cause and the proportionality of response to cause—the age-old distinction of normal sadness versus excessive sadness or sadness 'without cause'—were lost, with serious consequences for our own time, as we will soon see.

The situation in 1970 was a mess of theories and practices that had little or no consensus apart from agreement on the division between psychotic and neurotic depressions. Was depression of the non-psychotic type continuous with psychotic forms, or not? How many forms and of what sort were neurotic depres-sions? Did neurotic depression act as a stepping stone to the psychotic? Diagnosis was also haphazard, with little agreement on which symptoms attached to which neurotic depressions.

By the late 1970s, it was clear that something needed to be done, and the DSM-III was that 'something'.

Its forerunner, the *Diagnostic and Statistical Manual of Mental Disorders* (DSM-I), became the bedrock of American psychiatry in 1952, replacing the *Statistical Manual for the Use of Hospitals for Mental Diseases*, which had been the standard work of reference from 1918 to the early 1950s. The focus in the *Statistical Manual* had been on the more severe cases to be found in institutional settings rather than the milder neurotic ones that did not require hospitalisation. Psychiatrists worked largely in asylums, so naturally this was where their interests lay. By mid-century the emphasis in psychiatric practice had shifted outside the walls of the institution and now dealt more fully with less severe cases, using the psychodynamic tools largely bequeathed by psychoanalysis.

For this new scene, Meyer's work on reaction to various life circumstances and events (hence 'reactive depressions') and psychodynamic ideas concerning depression fed into the DSM-I and II. The stress fell on the talking cure, an attempt to reveal the unconscious conflicts that might be leading to depression, rather than the supposed biological causes postulated in the tradition of Griesinger and—to an extent—Kraepelin. Depressive disorders of the psychoneurotic variety were a result of the unconscious translating anxiety about loss into a different mode, according to the DSM-I. In true psychoanalytic fashion, it stated that 'the reaction is precipitated by a current situation, frequently by some loss sustained by the patient, and is often associated with a feeling of guilt for past failures or deeds'.[1] The 'intensity of the patient's ambivalent feeling toward his loss (love, possession)' in part determined the degree of the depressive reaction. Anxiety about a loss would prompt a defensive move by the patient's unconscious, and the degree of that anxiety would correspond to the degree of the depression. Psychotic depressions were a different category, and involved manic depression as well as other psychotic features that involved 'gross misinterpretation of reality, including, at times, delusions and hallucinations'.[2]

The DSM-II followed in the late 1960s, and defined 'neurotic' or 'reactive' (non-psychotic) depression as 'an excessive reaction of depression due to an internal conflict or to an identifiable event such as the loss of love object or cherished possession'.[3] It was not manic depression or involutional melancholia. Psychoanalysts were less concerned with specific symptoms than with the underlying psychodynamics that gave rise to either neurotic or psychotic depressions. This situation was to shift with the invention of a new form of depression via the paradigm-changing arrival of the DSM-III in 1980. In fact, in the early 1970s research psychiatrists at Washington University in St Louis were following in the tradition of Kraepelin (as they saw it) by trying to establish clear guidelines for research and diagnosis of depression and other disorders via statistical analysis of symptoms. Using a patient's symptoms as a diagnostic criterion, they would attempt to eliminate the ambiguity and downright confusion that was dogging their discipline.

This research was made more visible and coherent under the banner of John Feighner (1937–2006), an army physician turned into a psychiatrist by what he had seen of the psychological horrors of the Vietnam War and now also at the university, and who lent his name to the enormously influential *Feighner criteria*, describing the symptoms of fifteen psychological disorders. Feighner's career focused on treating mental illness through medications, and he made a name as a clinical pharmacologist;

the research institute he founded in 1972 conducted thousands of clinical trials. Although the team that produced them did not intend the diagnostic criteria for clinical use, they did think that they would provide common ground for researchers and establish clarity in comparison across different groups.[4] Like Kraepelin, they regarded depression as a unitary illness, placing all non-manic depressions into one category, and ignored research suggesting that there could be a distinction between psychotic unipolar depressions (that did not involve mania) and neurotic depressions.

Depression was a primary disorder of the emotions, and a diagnosis of depression required the patient to satisfy three criteria. The patient must have a dysphoric mood (be sad, feel hopeless etc.); must have five additional symptoms (lack of hunger, sleep, energy, interest in normal activities, speed of thought, suicidal thinking, guilt, agitation); and the symptoms must have lasted one month and not be caused by a different illness. All who conformed to these criteria were considered as depressive. The problem was that the Feighner criteria were not supported convincingly by existing studies, particularly in the collapse of the division between endogenous and reactive depressions. Nor did the criteria adequately provide any basis for distinguishing between patients who were normally sad (even beyond one month) and those who were excessively so.

Some of Feighner's colleagues at Washington University soon published the first textbook—*Psychiatric Diagnosis*—to rely on symptoms for diagnosis, with an emphasis on rising above debates about the causes of depression because of that very controversy. A 'cleaner' vision could be realised if speculation about etiology (the study of causes of disease) was avoided. Again, this book helped form the DSM-III, and continues in its sixth edition with new authors as *Goodwin and Guze's Psychiatric Diagnosis* (Oxford University Press, 2010), but still claims to be free of cumbersome theoretical matter.

The next building block in the production of the DSM-III was Robert Spitzer's (b.1932) *Research Diagnostic Criteria* (RDC) produced in 1978 in collaboration with Eli Robins (1921–95), a member of the Washington group and an opponent of the dominance of Freudian psychiatry in the 1940s. These were a translation of the Feighner criteria and were encouraged by the National Institute of Mental Health (NIMH, the largest research organisation in the world devoted to mental health, and established in 1949 by the US government) to solidify diagnostic reliability and to refine definitions of mental illness. Now there were twenty-five major forms of disorder. A significant change in these criteria was the shortened time limit: one month went down to two weeks, without explanation. Loss of interest could replace dysphoric mood, and the sufferer should exhibit social disability of some kind or have sought help. In short, the bar for a diagnosis of clinical depression was set even lower than in the Feighner criteria—there were no exclusions for normal reactions like death of a loved one, for example. Again, although the RDC were intended for research purposes and not others, the clinical application followed hard on their heels.

The RDC achieved the intended purpose: overall reliability of diagnosis was greatly improved within their own narrow and dubious definitions. It was all very well to have consistency of diagnosis, but was the actual judgement valid? 'These studies did not assess the validity of the diagnosis in predicting course, response to therapies, or etiology of depressive conditions.'[5] The problem of normal versus clinical sadness was also not adequately addressed, and remains a contentious matter.

Defining the new depression: the DSM-III and IV

Robert Spitzer's role as chair of the DSM-III task force gave him the chance to modify elements of the RDC and Feighner criteria, and create an approach to diagnosis that based itself on symptoms. The reason for the success of his attempt was that such a system addressed various pressing problems that were facing the psychiatric profession by the 1970s. The first problem was the decline of psychoanalysis and the concomitant rise of competing theoretical approaches. The new manual was theory neutral as it relied almost entirely on symptoms, and so nicely sidestepped the issue of theoretical persuasion. The emphasis now was on description rather than origin, thus causing the least offence to different schools of thought.

From a different direction, the anti-psychiatry movement, with psychiatrist Thomas Szasz (b.1920) leading the charge and ably reinforced by French theoretician Michel Foucault (1926–84) and R. D. Laing (1927–89) in the UK, alleged that psychiatry was merely a mode of social control rather than a genuine attempt to heal the ill. Although the anti-psychiatric movement became associated with the counter-culture of the 1960s, it had in fact been developing throughout the twentieth century. Ken Kesey's novel *One Flew Over the Cuckoo's Nest* (1962) and the subsequent academy award-winning film expressed fears about the abuse of institutional psychiatric power in the service of an oppressively conformist society. Behaviourists claimed that no mental disorder really existed, and that all behaviour was learned. On a practical level, by the 1980s the cost of medical treatment was being directed largely through insurance policies via either private or public bodies. These new diagnoses were apparently more accurate in their pinpointing of specific disorders to be paid for, and so suited the needs of insurers better. Obscure psychoanalytic explanations of a patient's condition were not as easy to process as a 'tick list' of symptoms.

Finally, the lack of agreement in diagnosis—the DSM-II lacked symptoms to determine specific diagnoses—undermined the psychiatric profession. The US–UK Diagnostic Project of 1972 found that the British diagnosed depression five times as much as the Americans—a clear embarrassment when the study was published. Even within cultures and nations the same presentation of a patient elicited an alarming variety of responses from psychiatrists, and it was not even clear that they could identify normal individuals from psychotic ones. How could the profession claim to be scientific if there was no agreement on who was sane or insane? If diagnoses could be more in agreement, then the psychiatrist and the profession would at least seem to be more trustworthy.

In one way, the DSM-III, as promoted by Spitzer, was a tremendous success because it helped alter the image of the psychiatric profession, and helpfully 'answered' the problems surrounding the unreliability of diagnosis, theoretical confusion, and anti-psychiatry attacks on the profession. In another way, it failed insofar as it ignored the question of the validity of the new system of symptom-based diagnosis. The DSM-III provided a new definition of depression, but, as we shall see, not necessarily a good one in itself and in its effects on treatment. Now depression was defined in terms of symptoms and not etiology, as in the Feighner criteria and RDC. Like them, external circumstances, apart from bereavement, were not considered as significant to diagnosis. Major depressive disorder (MDD), a 'unipolar' disorder, remained separate from manic depression, a 'bipolar' disorder (versus Kraepelin) and, though it encompassed psychotic depressions, focused largely on 'simple' depression, which was thought to be by far the most common form. Loaded terms and concepts like 'neurotic' and 'psychotic', endogenous and reactive or exogenous were jettisoned as part of the purging of theoretical constructs.

The lack of distinction between normal sadness and disordered mental states of depression in the DSM-III has had real and serious consequences. Treating people who are experiencing normal sadness responses to a range of life situations not considered in the DSM-III, such as divorce, illness, and money problems, has led to 'a massive pathologization of normal sadness that, ironically, can be argued to have made depressive diagnosis less rather than more scientifically valid'.[6] According to this argument, powerful antidepressant drugs with a variety of side effects that should be used sparingly among a seriously ill population have been extended into the general community, with the consequence that many people have been made worse rather than better by psychiatric intervention.

The DSM-IV-TR (2000), the 'text-revised' fourth edition, brings this story up to the first decade of the twenty-first century. There is little difference between the criteria in the third and fourth editions. Again, Major Depressive Disorder (MDD) is a unipolar disorder of mood, usually occurring in intervals of Major Depressive Episodes (MDE), and to be distinguished from Dysthymia (a lower-level form of depression), which is continuous for long time periods and is less intense than MDD. In effect, the diagnosis of MDD boils down to the presence of an MDE, which requires at least five defined symptoms (at least one of which should be loss of interest or pleasure) during a fortnightly period, occurring nearly every day and that are a shift from the previous state. The allowable symptoms include depressed mood, weight loss, insomnia, psychomotor agitation or retardation, fatigue, excessive guilt or feelings of worthlessness, reduction in ability to concentrate or make decisions, and recurrent thoughts of death or suicide. If symptoms are deemed to be clinically significant or affect the functioning of the individual markedly (at work or otherwise) then MDD can be diagnosed. Exclusions include bereavement (unless the depression continues over two months) and drug abuse. Again, it should not be a 'Mixed' or 'Manic' episode.

Objections to the present method of assessing depression include arguments that 'unusually harsh environmental stressors often produce many intense symptoms in otherwise normal individuals, and the depressive symptoms that occur during normal periods of sadness are generally similar to the depressive symptoms listed in the DSM criteria that occur during depressive disorders'.[7] Some people are naturally more sensitive to stress, and clearly normal reactions to major life events such as divorce or financial problems can easily last more than two weeks in any event.

The DSM-IV introduced a new 'subthreshold' category called Minor Depressive Disorder that would only need two symptoms from the nine MDD criteria, one being loss of pleasure or interest. As yet it has not been included as an official category, but clearly it is even more difficult in this case to distinguish between normal and pathological sadness. A further category is Dysthymic Disorder, previously known as the traditional category of neurotic depression or 'Depressive Neurosis' in the DSM-III, a concession to psychodynamic practitioners, requiring that the symptoms (mood disturbance and two others) must have persisted for two years on most days, but again without taking into account chronic stressors such as the slow decline and death of a child or close relative. This milder and chronic form of depressive condition has all the problems associated with the MDD in terms of the inability to distinguish between normal and abnormal sadness.

The interesting subcategory of 'Major Melancholic Depressive Disorder' is designed to echo traditional endogenous (internally caused) depression, although the etiology underlying that category (physiological disorder and not external triggers) is avoided and replaced, as usual, by symptom-based diagnosis. Here, the person has lost pleasure in nearly everything and has three symptoms from a list including distinctly depressed mood (worse in the morning), waking early, sluggishness, excessive weight loss, and guilt. The usual difficulty associated with the symptom-based approach of DSM-III and IV persists.

A major shift from traditional conceptions of melancholy and depression in the DSM-III is the attempt to conceive of a purely depressive syndrome that largely excludes anxiety. 'Fear and sadness' have been the two crucial elements of depression's biography since its classical 'birth', but here one half of the equation is lost in Major Depressive Disorder, despite the reality of the co-presence of the two in the majority of cases: they go hand-in-glove, however the emphasis falls. In the psychoanalytic perspective, of course, anxiety about loss produces depression.

DSM-III and IV provided a model by which straightforward checklist questionnaires could be administered by lower-level healthworkers or administrators, rather than demanding the time of a highly-paid specialist to diagnose every case of depression or other mental illness. The brave new world of mental healthcare could, theoretically, be extended to a much wider community outside the asylum than previously. In this sense the DSM-III was a tremendously useful tool and ticked many boxes in terms of the new post-war demands in the administration and funding of healthcare. Over time it would appear that the convenience of mass diagnosis by questionnaire has been outweighed by the decontextualisation of the symptoms and that the result has been a 'surveillance of sadness' that has encroached massively on normal sadness and pathologised far more people than is justified or necessary.

Competing biological models

It may or may not be a coincidence that biological models of depression began to dominate psychiatry at roughly the same time as the publication of the DSM-III in the 1980s. After all, the DSM-III was intended to bypass matters of theory and causation, but its criteria have aided the growth of research in the biology of depression by providing a level playing field on which all researchers communicate. Biological research into depression had been ongoing throughout the twentieth century, a continuation of older theories of hereditariness and brain lesions that we saw in previous chapters. Twin and adoption studies strove to show the genetic basis for depression throughout the twentieth century, with findings that have been murky. It appears that the heritability of depression can be around thirty to forty per cent, but it is not clear from such studies whether the transmitted element is normal sadness or depressive disorder.

Studies into depression in the new age of the 1990s, in which specific genes could be examined in relation to their connection to certain illnesses or symptoms, seem more promising, yet are still fraught with difficulties. One well-known study of the effect of the 5-HTT gene on depression did find that stressful life events correlated with depression, and further claimed that the short allele on this particular gene put the possessors at a higher risk by making them more sensitive to stress and therefore depression. These findings have been challenged, however: 'it is not clear that the identified gene has much to do with depressive disorder at all'.[8] A close examination of the study argues that social class might well be the determining factor for the results, and that attention might be better focused on alleviating environmental factors like poverty rather than a presumed genetic defect that might or might not be treatable by drugs (which is what the authors of the study proposed). More recent studies acknowledge the existence of environmental factors in a more sophisticated way, but are still—perhaps inevitably—hunting for the genetic culprits.[9] So far genetic research has not proved to be a solution to the treatment or even prevention of depression.

One of the more common biological and popular ideas about the causation of depression is that it stems from a chemical (no longer humoral) imbalance. Levels of chemicals in the body are studied and related to the presence of depressive disorders. An early candidate was norepinephrine, one of the amine family, as proposed by Joseph Schildkraut in 1965: too little meant depression, too much meant elation. Serotonin deficiency is possibly the most well-known of the contenders for causing depression.

More recently, it has been thought that other processes in changing amine activity are involved in depression, as SSRIs (selective serotonin reuptake inhibitors) change serotonin levels immediately, yet take weeks to work on the depression. Furthermore, some drugs help reduce depression yet do not work on either serotonin or norepinephrine, and some affect dopamine and other amines, but not serotonin. Also, some antidepressant drugs are effective in treating other illnesses like substance abuse, attention deficit and anxiety, which implies that drugs act generally on brain function rather than specifically on depression. Apparently, a mere quarter of depressives have low levels of serotonin or norepinephrine although, as with so much associated with depression, this figure is controversial, to say the least.[10]

To add to the problems with the chemical imbalance theory, it is not clear whether the alleged deficiencies in brain chemicals are a result of depression, rather than a (preceding) cause. How, moreover, do we know what level of serotonin, or other neurotransmitter, is normal, when neurochemical levels vary contextually and naturally according to particular situations? They are responsive to different situations and within different individual constitutions. Changes in serotonin levels could reflect shifting and stressful life events rather than being evidence of depressive illness. Even in the apparently clear case of post-natal depression, where it seems as if hormonal changes are the cause, there has not been sufficient research into the range of factors (such as the family history of depression) that might contribute to its onset.

These are a few of the many issues associated with chemical imbalance theory, but it is safe to say that the exact processes through which some drugs might alleviate depression and other mental illness are not sufficiently well understood.

A further popular form of biological investigation is the quest to find abnormalities in the brain that would cause depression. Different areas of the brain have different tasks: the prefrontal cortex helps to control changes in mood and anxiety; the hippocampus involves learning and memory; the amygdala handles

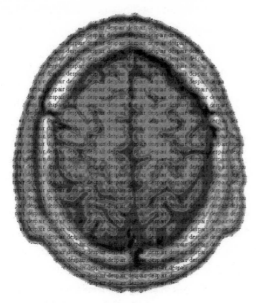

Fig. 15 MRI scan of the brain overlaid with 'despair', by Mark Lythgoe. His work focuses on the brain and here invests it with an emotional meaning that the original purely scientific image cannot. (*Mark Lythgoe, Wellcome Library, London*)

negative emotions. The use of Magnetic Resonance Imaging (MRI) scanners in particular has opened up the possibility that the brain can be scrutinised at a level of detail which might enable the specific lesions that supposedly cause depression to be identified (see Fig. 15). The problem with such research at present is that it is unclear whether depression causes the brain damage, or the brain damage causes the depression. It might in fact be the antidepressant medication of the patients in the studies that caused the brain lesions. The healthy brain of the same patient would need to be scanned in order to have a true comparison of the changes in brain structure. Modifications in the relevant regions of the brain can be caused by low mood in both normal and pathological individuals.

Somatic treatments and the rise of antidepressant drugs

The history of drug use in the mid-twentieth century, at least in America, is dominated by anti-anxiety treatments like the tranquilliser meprobamate (Miltown) in the 1950s, and then in the 1960s the benzodiazepines Librium and Valium, which were normally prescribed by the general practitioner. The focus here was to alleviate the stress of everyday living, with one in twenty Americans taking some form of sedative by 1956. The 'stress tradition', with its concomitant annexation of anxiety (once hand-in-glove with depression), emerged partly from psychoanalytic ideas but also from Meyerian notions of environmental factors on mental health, and war psychiatry that demonstrated the effects of stress in extreme circumstances on even the strongest individual. Normal sadness, or non-psychotic forms of depression, tended to be folded into a diagnosis of underlying anxiety before the 1980s, despite a rising interest in depression in primary care from researchers as early as the 1960s.

Drugs for the specific treatment of depression also appeared mid-century, such as the monoamine oxidase inhibitors (MAOIs) and tricyclic antidepressants, such as imipramine and amitriptyline. These were prescribed in substantial numbers, but nowhere near as many as for anxiety. These antidepressant drugs also had substantial side effects that curbed their use for a wider population. In that period, before the publication of the DSM-III, depression was seen as a limited problem, as opposed to the 'age of anxiety', as W. H. Auden dubbed it in his poem of that title (1947), in which life after the war was characterised by multiple social and political fears.

Between the 1930s and 1950s the most widely used somatic treatments for depression and other mental illnesses were the now-controversial electroconvulsive therapy (ECT) and lobotomy. Severe depression seemed best treated by Metrazol in combination with electric shock, and ECT continues to have limited application to this day. Edward Shorter claims that ECT 'remains the most effective treatment of serious, melancholic depression', despite the tremendous stigma attached to its use through its previous actual misuse and image of ECT as a tool for social control, most famously in *One Flew Over the Cuckoo's Nest*.[11] Modern treatment requires informed consent in all but the most serious cases in the UK and the United States; risks remain, however, and patient experiences are far from universally positive. Lobotomy—cutting the links between the prefrontal cortex (part of the frontal lobes of the brain) and the rest of the brain—largely fell out of favour by the early 1970s and, unlike ECT, has not been rehabilitated. It has been seen as one of the so-called 'heroic' treatments of the pre-antidepressant era, when psychiatrists were keen to develop effective somatic treatments.

In the late 1960s there was a backlash against the numbing of people's reactions to normal life by over-prescription of anxiolytics (anti-anxiety drugs). Popular and medical opinion shifted in favour of restricting the use of 'Mother's Little Helper' (The Rolling Stones) and limits were introduced on the claims made by advertisers about treating the everyday strains and stresses. There were also fears about the dangers of these drugs in terms of side effects and addiction.

The arrival of the DSM-III proved fortuitous for the pharmaceutical industry. Because American regulations since 1962 had forbidden the targeting of drugs at generalised stress rather than specific diseases, the pinning of exact categorical diseases to symptoms in the DSM-III allowed the targeting of drugs at these new and numerous conditions. Major Depressive Disorder, which previously might have been seen as a function of everyday stress, now became a prime market for pharmaceutical companies. The shift away from anxiety drugs was aided by the new attempt to isolate major depression as a disorder not characterised by anxiety.

Moreover, the new research into serotonin and the use of SSRIs to raise its levels in the brain helped the expansion of treatments for depression in the 1980s. The new drugs were allegedly safer than the old ones and required less medical supervision. SSRIs affect the overall functioning of the brain in both normal and ill subjects, and have subsequently been used to treat other disorders or problems, notably anxiety and related disorders such as obsessive–compulsive, eating, and attention-deficit disorder. When the SSRIs were approved for use in the late 1980s they were registered only to treat depression, despite their wider possibilities. This avoidance of the negative implications of anxiety drugs (their side effects and particularly fears about addiction) gave a greater drive to push depression as the disease to market and the disease to cure with the allegedly wonderful new SSRIs.

The stage was set for Prozac, or fluoxetine, to dominate the antidepressant drug scene for the rest of the century: Peter Kramer's *Listening to Prozac: A Psychiatrist Explores Antidepressant Drugs and the Remaking of the Self* (1993) mythologised the whole class of SSRIs as a way of improving the lives of those both depressed and normal. Prozac allegedly gave energy and inspiration, whereas the tranquillisers deadened and dulled life for the consumer. Here was a new wonder drug, whose side effects were not yet known: the magic bullet has an allure in all ages, and Prozac fitted the bill for the last part of the twentieth century. It was the second best-selling drug in the world by 1994, numerical evidence of an accompanying cultural ascendancy that was to be reflected and fuelled in popular memoirs such as Elizabeth Wurtzel's *Prozac Nation: Young and Depressed in America* (1994). Wurtzel's fame also helped confirm depression in the public imagination as a primarily female disorder, as well as one that seemed newly a disease of the teenager (which was not very significant statistically in comparison to depression and suicide in the aged). *Prozac Nation* enabled depression to be newly fashionable in a similar way to Goethe's novel about the character of young Werther in the eighteenth century.

The idea that Prozac and its relations in the SSRI family could act as 'enhancement technology' and make people 'better than well' persisted throughout the 1990s and into the new century, not least because some people did experience an apparently miraculous transformation from depressive to 'normal', or even better than normal. Lauren Slater's *Prozac Diary* (Random House, 1998) initially describes the total turnaround of her life thanks to Prozac, although the longer the story goes on, the more the side effects of the drug begin to tell. Slater started taking the drug in 1988, so the trajectory of her diary in a way mirrors the fortunes of Prozac itself into the twenty-first century, where increasing doubts about its long-term effects and indeed its effectiveness in general have grown.

Various institutional factors aided the growth of Prozac in both the United States and Britain: the general practitioner supplanted the psychiatrist as the primary prescriber of antidepressants (as opposed to the anxiolytics of the 1950s and 1960s). A ten-minute appointment with a general practitioner can result in the prescription of Prozac for a range of disorders, not merely depression. The short-term cost of this form of healthcare is low in relative terms, and such prescriptions have the advantage of being efficient to deliver for health insurance companies and care plans. Consumers also began to demand the drug themselves, especially after the approval of direct marketing to consumers by the Federal Drug Administration in 1997. Thanks to the DSM-III, drug companies were able to claim legitimately that they were targeting specific conditions, even if one drug might serve many purposes, as did Prozac. Also, the DSM's wide range of common symptoms for Major Depressive Disorder drew in a wider population who might be defined as clinically depressed, and who could now regard themselves as depressed, thus expanding the market for drugs and depression exponentially.

The promotion of drugs to treat the 'wealthy well' and the expansion of the healthcare market to those now defined as depressed by the DSM-III and IV is a highly contested tale. It is clear that the role of pharmaceutical companies in funding medical research in and outside the academy, promoting screening for depression and other conditions, political lobbying, and particularly their funding of and control over the results of drug trials, has been a major factor in the way depression is defined and treated in recent decades. There is now a widespread acceptance of the routine prescription of drugs like Prozac, even if a backlash similar to that experienced with the anti-anxiety drugs is now occurring, with a plethora of books railing against the activities of drug companies in expanding their market for antidepressants.

We now approach the thorny debate about the effectiveness or otherwise of Prozac and other antidepressants. This is a difficult area to assess: the pro- and anti-Prozac camps both enlist 'science' to prove their own case, and use powerful rhetoric to discuss the way in which drug trials are conducted and statistical conclusions reached.

Proponents of medication for depression argue that drugs have undoubtedly been useful in the treatment of MDD, and, if nothing else, can buy time to stabilise the patient. In this they are supported by the admittedly ambiguous anecdotal evidence in the various literary and autobiographical narratives describing the apparently miraculous effects of Prozac in particular. As Ronald Wallace put it in 'On Prozac':

> So much happiness! It seems
> everything I touch shines back, all smiles.[12]

Wallace, or at least his poetic character, does qualify this chemical joy with the vital question of identity, however: 'But is this what I wanted, after all?'

An *un*ambiguous endorsement of the use of antidepressant drugs is to be found in official guidelines, in which Prozac is still the first means of treating and preventing more serious depression recommended to psychiatrists and family doctors in America. Although the rise of Cognitive Behavioural Therapy has made inroads into medical practice and public consciousness, access to such 'talking' therapies remains erratic and is perceived to be expensive. Psychoanalysis plays a very small role in the spectrum of treatments available via public health services. Despite the intensifying critique of SSRIs and their side effects, the official medical perception remains that these drugs have few side effects and provide effective relief for the depressed patient.

More than this, some studies show that there is a need for more medication of depressives. According to this perspective, if more people realise that they are depressed they could seek the readily available cure for their condition. To Gerald Klerman, it is merely 'psychological Calvinism' to argue that it is 'weak' to take drugs to cure a medical condition—this is a simple question of treatment, not morality. Why take the alleged moral high ground by going through expensive and lengthy psychoanalysis when there is a straightforward cure?[13] The logical extension of this attitude is that there is no harm in using Prozac as a 'lifestyle' drug if it adds to the quality of one's life.

The most strident criticism of antidepressant drugs comes in the work of David Healy, Professor of Psychiatry at Cardiff University, and author of *The Anti-Depressant Era* (Harvard University Press, 1997) and *Let Them Eat Prozac: The Unhealthy Relationship Between the Pharmaceutical Industry and Depression* (2004), in which he has alleged that the use of Prozac increases the risk of suicide in younger patients especially. More recently, he describes a bipolar patient he calls 'Alex' who dies while being treated with an antipsychotic drug. Alex was two years old.[14] As a professional 'insider', Healy's work has tended to carry more weight, and the fact that he is not against the use of medication per se gives his claims further validity. Healy's ideas are reinforced by an increasing number of other authors who write for a wider market than the purely academic: Irving Kirsch's *The Emperor's New Drugs: Exploding the Antidepressant Myth* (Bodley Head, 2009), Richard Bentall's *Doctoring the Mind: Why Psychiatric Treatments Fail* (Allen Lane, 2009), and Gary Greenberg's *Manufacturing Depression: The Secret History of a Modern Disease* (Bloomsbury, 2010).

Healy's school of criticism is concerned with the dangers posed by specific classes of drugs and the way in which the interests of drug companies are allegedly allowed to flourish due to their high degree of control over the production of (accessible) knowledge about antidepressants and, partly, the 'invention' and categorisation of the diseases that they are designed to treat. According to Healy and the medical historian Edward Shorter, the effectiveness of SSRIs has been greatly exaggerated: they are not necessarily more curative than placebo. There is a large debate on the nature and extent of the 'placebo effect' beyond the treatment of depression, but it has a particular potency in this context. The side effects of the SSRIs are also more serious and common than has previously been allowed with headaches, nausea, stomach problems, and loss of libido leading a long list. More profoundly, Healy has stated that the rise of depression as a particular disease was due to the introduction of drugs for mental illness in the 1950s in a hospital context, a theory which we will see is shared in various ways by other theorists of the 'new' depression.

Some drugs, moreover, are unattractive for marketing purposes because they have become unprofitable. Edward Shorter claims that existing drugs (barbiturates, opiates, and amphetamines) are better at treating depression and have fewer side effects than the newer SSRIs, with the caveat that they are prescribed only where there is careful supervision by trained physicians.[15] Expired patents do not generate fresh cash for pharmaceutical companies, so it is in the companies' interests to push new drug classes.

The final word here on drug effectiveness should go to the patients themselves. The mental health charity Mind has published a book examining the experience of psychiatric drug users, an experiential perspective not often considered in the vast biomedical literature on depression. While it is true that such impressionist accounts can be problematic, Jim Read, the book's author and a former user of psychiatric medication himself, has found that the patient experience is extremely variable, with some people claiming that the drugs do work and experiencing few side effects, and others having their lives ruined by the side effects. The great difference in individual responses to psychiatric medications has meant that a certain amount of experimentation to find the best, or least worst, forms of drug is common. The ingestion of a drug into the human body is a chemical process, but the way it is perceived and experienced depends on the individual's social and personal values and perspectives. If, for example, a person expects the experience of Prozac to be fashionably creative, then this might become a self-fulfilling prophecy (even if the chemical effects of the drug are negative). However, some specific drugs within broad classes (like SSRIs) do have more side effects in general than others, and some drug classes are more problematic than others.[16]

The 'drug' question leads us on to the related problem of the explosion in the diagnosis and treatment of depression at the end of the twentieth century. One explanation for this expansion is that Western capitalism, at least in its more recent forms, produces depression: depression is a result of the pressures of modern life, much as neurasthenia was thought to be in the nineteenth century. This does beg the question of class and gender within this hypothesised capitalist lifestyle. Are the experiences of manual workers, for example, to be compared with those of the middle-class office worker? We might expect poverty to be a driver for depression, and we have seen this debate expressed throughout the history of melancholia and depression.

Gender is a major problem in the dispute about treating depression with drugs, especially given that women are constantly depicted as being the largest group of depressed patients. Is it true that women are really more depressed than men? Was gender was a central factor in this patient demand, and is 'Big

Pharma' more than a mere pantomime villain? Having already been encouraged to see themselves as active consumers of medical services, especially in the second part of the twentieth century, women were ready to move on to describing themselves as depressed when drugs specific to its 'cure' were introduced in the 1980s. The more macho discourse of masculinity in capitalist culture fed differently into the consumption of medical services, especially in relation to depression, and made it more difficult for men to see themselves in this new disease. The 'language of feelings' so amenable to feminine discourse was well suited to the language of DSM-III symptoms, which in turn had been constructed on samples of depressive patients, many of whom were women. John Feighner's crucial work on the diagnostic criteria with his group at Washington University, for example, used studies such as one in which there were sixteen men and thirty-three women, explaining such a gender bias by arguing in a circular manner that it reflected the typical prevalence of women in diagnosed depression. It is unsurprising, therefore, that the DSM-III produced a form of depression that 'fitted' women. 'Irritability' and mysterious physical symptoms were dropped, both of which might have drawn in more men.

Contemporary campaigns aimed at educating men to understand when they may be depressed have tended to stress depression as a problem of biochemistry, 'a disease not a feeling', whereas those targeting women have encouraged women to regard their feelings of low mood as a disease. Both shared the assumption that this disease, however differentially framed for each gender, is treatable by a pill. Women's distress, which might well have clear social causes, became validated by the broad and female-friendly definition of depression. Depression's image as a female disease has paradoxically been welcomed by feminists otherwise highly critical of the psychiatric profession's construction of female maladies as a result of their peculiar biology, even though the new depression equally ascribed its existence to gendered physiological characteristics. Women have consumed depression and its cure eagerly because these take away personal responsibility and social stigma.

A contentious argument, partly informed by the postmodern ideas of French theorist Michel Foucault about the construction of our reality (or realities) by the power of different discourses, is that depression in its present form is not actually a disease in the strictest sense at all. This can be evidenced perhaps by the small number of cases diagnosed before the advent of the SSRIs.[17] The pharmaceutical company Geigy refused to fund the development of the antidepressant imipramine in the mid-1950s because it thought the market was insignificant—a contrast indeed with the massive sales of Prozac in the last part of the century. It is no coincidence that fin-de-siècle depression was 'invented' at the same time as the new drugs to treat it.

Yet this is not merely a case of drug companies pushing drugs and inventing diseases to 'match' the often scattergun effects of those drugs (something made much easier by the new DSM-III symptom-based definitions), but also because patients were not passive. Sufferers of genuine symptoms of lowness, insomnia, psychomotor retardation, and so on—whether of normal or pathological sadness—were able to see themselves in the new definition of depression described by the drug companies. Moreover, the new drugs did actually work to alleviate their symptoms, perhaps with a powerful placebo effect, and allegedly had fewer side effects (at least in the short term) than their predecessors. One's confidence in this fact might be lessened given more recent attacks on the data and associated image of the effectiveness of SSRIs, especially on normal sadness. Patient demand helped fuel the growth of the 'new' depression.

The postmodern perspective on the cultural construction of depression does not deny the existence of genuine symptoms of distress: 'melancholia', 'depressive psychosis' or 'endogenous depression' are 'described in remarkably consistent terms' from Hippocrates to the present day (Borch-Jacobsen, 198). Depression is not a myth or an illusion, 'but this reality is not hard-wired in...[the] genes or neurotransmitters. It was fabricated, constructed, produced, invented by the drugs of the bio-medical industry, with which it is of one piece. In that sense, it is not a fate: change the medication and the therapy, and we would have a new illness' (Borch-Jacobsen, 203). While provocatively expressed, such arguments remind us of the agency of the patient in the production of disease and its treatment. In much the same way, in the eighteenth-century medical market the wealthier patients were able to demand certain treatments for their (often self-diagnosed) melancholy and diseases of the nerves.

Worse still, are antidepressant drugs effective at all? Irving Kirsch's *The Emperor's New Drugs: Exploding the Antidepressant Myth* (2009) is a more extreme extension of the controversial critique of antidepressant drugs and the contemporary idea of depression as a disease. Kirsch is a clinical psychologist who has done research into the placebo effect at the University of Connecticut and become dissatisfied with the explanations of why the drugs prescribed by his psychiatric colleagues seemed to result in improvements in depressed patients. He argues that antidepressants are only marginally more effective than placebos. His 'meta-analysis' of a number of clinical trials (hidden by pharmaceutical companies and exposed by Kirsch via the Freedom of Information Act) revealed that even in severe depression the antidepressants were only slightly more effective than a placebo (33). Conditions that involve a greater element of subjective pain or suffering tend to be more amenable to the placebo effect, and even placebo surgery for osteoarthritic knees was found to be more effective than the actual 'scraping and rinsing' surgery in a 1990s study. He flatly dismisses the chemical imbalance theory, pointing out that some drugs like tianeptine are as effective as the SSRIs despite the fact that these new drugs reduce the amount of serotonin in the brain rather than increasing it. He announces that the paradigm of the chemical imbalance theory is over, and that therapies like Cognitive Behavioural Therapy (see later) are the solution. The new idea of neural plasticity is little better, allegedly. According to this theory, the brain can change when people absorb new information, so that depressives can be cured by treatments that teach them how to process information better, including talking therapies like Cognitive Behavioural Therapy. The difficulty with the theory of neural plasticity is knowing how all the different therapies, drugs or otherwise, might affect neural networks.

We end this chapter discussing the notion that antidepressants are dependant on the placebo effect and, in essence, do not work, because this critique is the most extreme position taken on the biological model that has dominated the last thirty years. Various parties, not all of them in thrall to the pharmaceutical industry, have roundly rebuffed this challenge. Nevertheless, attacking the effectiveness of all antidepressant drugs and the biomedical model that underpins their use suggests that something changed in the early years of the present century. The final chapter of this book will explore what that might be, and what other models might move the story of depression forward.

VII

'THE DRUGS DON'T WORK'?

The Future for Depression and Melancholia

We now live with a tension between the biochemical model of depression and the psychological conflicts that suggest other ways of viewing the illness. The 'Prozac' perspective has exerted a very strong hold over the public imagination in previous decades, yet there are signs of a shift in paradigm, as we saw at the end of the last chapter. The very idea of the widespread use of the DSM-III's 'new' depression is under scrutiny, but what alternative definitions and treatments exist and await? In this final chapter we pick up threads from the history of depression that go beyond the straitjacket of the biochemical model and have recently come to the fore as the potential of Prozac recedes.

Freud-free talking cures

The most significant non-somatic treatment for depression in the first decade of the twenty-first century has been Cognitive Behavioural Therapy (CBT). Aaron T. Beck (b.1921), Professor Emeritus in Psychiatry at the University of Pennsylvania, famously bucked the trends which labelled depression as a disorder of the emotions or of biology. For him depression was a result of cognitive error, mistakes in thought that were correctable—an echo of the eighteenth-century idea of melancholia. He began publishing on the subject in the late 1960s, but his approach has only comparatively recently become a common and frequently effective tool in the battle against depression, as recommended in WHO (World Health Organization) guidelines.

For Beck, depression followed a certain sequence and was caused by some kind of loss, at least as perceived by the depressive. Beck considered that different people assess loss in different ways. If one's wife runs away with another person, one person will grieve but then move on with his life; if he has strongly identified his entire persona with her, he might be plunged into deep depression, his world shattered. Thus, the event itself is framed differently and elicits different responses: it is not events in themselves that are depressing, but the meanings placed on them by particular individuals. So, the event of a wife being 'lost' might precipitate a highly negative cognitive state, which generates a chain reaction progressing to a full-blown depression. The person might question his sense of self-worth ('If I had been a better person she wouldn't have left me') and proceed to further pessimistic or catastrophic conclusions about his life.[1] For Beck, the husband at this point was making cognitive mistakes and if uncorrected they would result in the development of an unfortunate feedback mechanism. The physiological consequences of depression, such as sleeplessness and loss of appetite, exacerbate and reinforce his psychological condition. He concludes that 'I shall always be sad' and that he will never sleep again (and so on). Beck's therapy consisted of correcting the pessimism of the depressive by suggesting alternative, realistic thoughts—'I may be suffering at the moment but I can still live a good and enjoyable life eventually.' These can replace the negativity driving the depression. By breaking the vicious circle of depressive thinking the physical symptoms will lessen and a virtuous circle of positive or at least realistic thinking will ensue.

Beck was on weaker ground, argue some, when he stated that all depressions conform to his chronology of development, even those with no apparent cause. One strength, say others, was his liberation of the concept of loss from a Freudian underpinning: Beck was not bound by the psychoanalytical framework and the mother–child dyad. Loss for Beck was that which is defined as such by the patient, not the analyst. The entire technique of cognitive therapy has been criticised as a 'quick fix' (comparatively speaking) and suited to the 'repair' of individuals the better to return them to a condition ready to function in a capitalist society.

A more recent development of CBT is Mindfulness-Based Cognitive Therapy (MBCT), which uses Buddhist meditation techniques in order to focus on the present moment rather than worrying about the past or future, and accept thoughts and feelings without placing a value on them, thus hopefully neutralising the cognitive errors that drive depressive emotions. MCBT has received practical frontline endorsement in Britain via the National Institute of Clinical Excellence (NICE), which has recommended MBCT for patients that have suffered from three or more major episodes of depression, and individual CBT for people who have relapsed despite antidepressant medication or have a history of relapse. For people with lower-level or 'subthreshold' symptoms, computerised CBT is also recommended by NICE.

Learned helplessness

Closely related to CBT, developed at roughly the same time, and enormously cited in its own right, is Martin Seligman's well-established concept of 'learned helplessness', which he developed in the late 1960s at the University of Pennsylvania, partly inspired by Beck's work at the same institution. Seligman is one of the most important figures in American psychiatry, and has been president of the American Psychological Association (APA) as well as an assiduous promoter of the nascent discipline of positive psychology and author of many self-help books. Seligman's (b.1942) distinctive model of depression—derived from experiments on laboratory animals—explains it as an aspect of learned helplessness, in which the individual comes to believe that she or he has no power over events in their own lives—that they are helpless. 'The label "depression" applies to passive individuals who believe they cannot do anything to relieve their suffering, who become depressed when they lose an important source of nurture—the perfect case for learned helplessness to model.'[2] Seligman also believed that his model worked for agitated depressions, and where there is no obvious external cause ('endogenous' depressions). He saw a continuum of susceptibility to believing in one's helplessness that underlies the endogenous–reactive continuum.

Here his (rather stereotypical) example is of a woman who breaks a dish just before her period and sets off a full-blown depression along with feelings of helplessness. At other times of the month this event would not have produced this particular depressive reaction: 'it would take several successive major traumas for depression to set in'.[3] For the 'reactive' or 'exogenous' depressions, the causes emanate from life experiences, like 'failure at work and school, the death of a loved one, rejection or separation from friends and loved ones, physical disease, financial difficulty, being faced with insoluble problems, and growing old'.[4]

The symptoms of learned helplessness parallel those of depression, at least as defined in this theory: lowered initiation of voluntary responses; negative state of mind; persistence of helplessness over time; lowered aggression; loss of appetite and libido; physiological changes (helpless rats show norepinephrine depletion). Problematically, Seligman's model makes it possible to eliminate aspects of depression and melancholia previously regarded as indispensable: sadness, a subjective state not measurable in the laboratory, is not a necessary component of depression in his terms. A related problem is that of loss in his theory, because the loss he describes seems more to do with control of one's environment and destiny than any deeper psychological concept.

Seligman's position on the debate as to whether depression is a disorder of mood or thought, affective or cognitive, is subtle. He argued that the distinction between the two was 'untenable' and merely an accident of our language. In reality, 'cognitions of helplessness lower mood, and a lowered mood, which may be brought about physiologically, increases susceptibility to cognitions of helplessness; indeed, this is the most insidious vicious circle in depression'.[5] In this, Seligman endorsed the treatment of depression by Cognitive Behavioural Therapy as a way of correcting the mindset and therefore the mood.

Control was a central theme for Seligman: he speculated that those who have had an early experience with uncontrollable events might be more prone to depression. Treatment requires the patient 'regaining his belief that he can control events important to him'.[6] It has been pointed out that one side effect of Seligman's thinking is its usefulness for feminist and race theory: oppressed groups in any society are, according to this model, prime candidates for depression via learned helplessness.

Freudian futures: psychoanalysis into the twenty-first century

From the psychoanalytical point of view, CBT and the idea of learned helplessness do not properly address the long-term causes of unconscious psychic conflict that underlie depression. CBT works to an extent, and keeps health managers happy, but is essentially superficial from this perspective. Psychoanalysis, on the other hand, has no guaranteed outcome because it is an open-ended process in which the analyst does not claim superior knowledge about the patient. Working through the patient's problems during psychoanalysis may provide a better long-term fix, but such an approach does not sit well with the demands of healthcare systems that desire more certainty and a quicker result than psychoanalysis can provide.

Psychoanalysis in most of the Western world has been marginalised by the biomedical model since the DSM-III, and has carried less credit as a treatment in the new world of evidence-based medicine.[7] The discipline has been heavily criticised in both its methodology and practice. In terms of practical delivery, it is covered to an extent by health insurance in Germany and France (where the Freudian tradition is strong in both hospital and outpatient contexts, unsurprisingly); there is very limited availability in the British National Health Service. In twentieth-century America psychoanalysis was generally privately funded unless the patient was residing in a publicly funded asylum, but in 1967 the Federal Employees Health Benefits Program started to give generous coverage to outpatient psychodynamic therapies, including psychoanalysis, coverage that was subsequently reduced when the costs of such a service became clear. The recent history of state provision of psychoanalysis has been one of cost containment, however, especially in the light of cheaper drug therapies and other forms of psychodynamic therapy. It remains to be seen whether the faith placed in psychoanalysis before the 'drug revolution' can be restored, and whether newer, more gender-, race- and class-aware models can make a difference to patients.

Depression and society

Sociological studies have done the great service of reaffirming the significance of social factors and life events in the onset of depression. Perhaps the foremost exemplar of the sociological approach is George Brown. This British sociologist and psychologist, described as 'possibly the world's pre-eminent researcher on depression', has had a great influence on psychiatry.[8] His 1978 book with Tirril Harris on the *Social origins of depression: A study of psychiatric disorder in women* demonstrated that only four of thirty-seven clinically depressed women in south London had causes of depression not linked to life events such as bad relationships and bereavements.[9] Only a small number of his sample had depression due to some external cause and not primarily an alleged inbuilt personality defect (of whatever cause). This might not seem much of a surprise given what we have seen of the early history of depression and melancholia, but Brown's work provided an evidence base for an alternative view

of depression from the biomedical or narrowly Freudian explanations that persisted throughout the twentieth century.

If we understand the causes of depression in society, argued
Brown, we can move towards prevention as well as cure. He
developed the Present State Exam (PSE) in the early 1970s, a
measurement of depression that has dovetailed with the DSM-
III in the sense that both tend to provide similar diagnoses,
although Brown's assessment only examines the previous year,
rather than the life history as in the DSM. Changing the social
stressors, according to Brown, can change the depression. He
particularly regarded humiliating (subordinating or undermining self-esteem) or entrapping (preventing escape from a situation) losses to be causes of subsequent depression. Almost fifty
per cent of women who had experienced these sorts of losses
were three times more likely to develop depression than ones
experiencing an isolated loss event. The picture is complicated
by Brown's observation that such losses are deeply bound up
with the self-image and aspirations of the individual, with personal psychology. Brown's cross-cultural studies found that
there were wide variations in rates of depression in women (and
men) across different societies, and that psychosocial issues
predominate, therefore, over biological ones.

Race and ethnicity also play their part in both the diagnosis
and the treatment of depression. In an interview with a group
of African and Asian heritage psychiatric hospital patients in
the UK in 2000 for the mental health charity Mind, 'R' observed
that 'Even with drugs, they don't say "let's put you on antidepressants", they don't recognise depression in black people. It's
either an antipsychotic or a mood stabiliser.'[10] Patients in the
Black Service User Group complained that they were perceived
as 'aggressive' and 'rebellious' rather than depressed because

'we've got attitude about us'. The Mind survey showed that black service users had less choice and less explanation than their white counterparts.

Suicide has also been argued to be primarily motivated by social rather than biological factors. Studies have shown that sociocultural factors are important determinants of self-harm, and that more effective treatments for depression do not correlate with a reduction in suicides. The group most at risk of suicide in Britain and the United States are Caucasian males over the age of 70, and people over 65 in general are more likely to commit suicide than younger age groups, despite the sporadic media focus on suicide in teenagers. Suicide rates in older men were found to be higher and rising in America in past decades than in the United Kingdom, where they actually decreased. Primary care interventions and suicide prevention centres appear to have made little difference. Access to *means* of suicide, however, such as higher rates of handgun ownership in the United States, have been found to be perhaps obvious but more relevant factors than the treatment of depression. Some critics have argued for a revision of primary care and its treatment of depression so that on-the-ground effectiveness is the goal, rather than stemming from a biomedical top-down model led by hospital psychiatry.

Darwin and depression: evolutionary theories

Another recent approach to depression seems, on the face of it, to be at odds with sociological interpretations of depression, and more related to the biomedical discourse described in Chapter 5. The trend to view psychology from an evolutionary perspective has yielded some interesting, if unproven, theories about the nature of depression, and why sadness exists at all. From the dawn of the human race, say evolutionary psychologists, depression has been a feature of our experience.

A range of ideas about the evolutionary value of depression includes the apparently paradoxical thought that depression can actually be useful to humans. On the face of it, depression interferes with reproduction, productivity, and even appetite. Here the distinction between normal and pathological sadness and depression comes in useful: some sadness responses must have been functional, so one argument goes, at an earlier evolutionary point. They were adaptive, and therefore normal. One hypothesis has it that human genetic traits were shaped in the 'environment of evolutionary adaptation' (EEA) when hunter-gatherers were on the African plains between ten thousand and two million years ago. Unfortunately, sadness responses that worked then might not be well adapted to the present environment of 'late' capitalism: it is a long way from the cave (or plain) to the office.

Depression may be a response to losing status as it evolved from the competition for pack primacy in the animal world. If an animal is beaten by another one, depression results as a way of preventing further damage, thus adapting to a situation in which subordination is the only course of action. This 'alpha-male' scenario seems suited to cut-throat capitalism where the concrete jungle requires the survival of the fittest, but this theory also has the advantage of explaining why subordinate groups, who are forced to inhibit their self-expression, might be subject to depression. Although we have contested this 'fact' in the previous chapter, the apparently higher rates of depression in women are explicable presumably because their social situation is often worse than that of men, while certain ethnic groups might also be in the same position, depending on their particular socio-economic context.

Randoph Nesse, influential Professor of Psychology and Psychiatry in Michigan, has pushed the evolutionary agenda from the 1990s onwards, and has proposed that a sadness response can prevent the individual from worsening a difficult or critical situation by preventing rash actions and decisions. The loss of energy that comes with sadness can have a protective effect. Such a normal sadness adaptation to a serious life situation can lead to a period in which the person's life can be reassessed and a more positive outcome emerge over time. When this response becomes disordered, rather than being a breathing space, the depression deepens and does not result in repair. Similarly, unreachable or very difficult goals might meet with a depressive response, an adaptive action designed to extricate the individual from that unproductive situation. By this withdrawal, the person is eventually freed to reinvest that energy in more achievable targets.

Evolutionary psychology has even taken on the perennial question of depression's connection with creativity, a long tradition that even recent literary works persist in reaffirming. The enforced suffering of depression can build empathy, humility, compassion, which might have assisted the survival in evolutionary terms of those individuals so afflicted. Despite the many studies counting the heads of depressive artists and writers, politicians and scientists, the relationship of depression to genius remains the subject of speculation, although it does seem likely that depression promotes a certain introspection, a different and potentially artistic understanding of reality. As the Romantic poet William Cowper put it: 'Dejection of spirits, which I suppose may have prevented many a man from becoming an author, made me one.'[11]

Cross-cultural psychiatry: is depression universal?

The Japanese character for depression in early modern times, 'utsusho', is composed of two characters, the first of which resembles a dense, dark and seemingly impenetrable thicket of strokes that symbolise the barrier between depressives and their grasp of hope and happiness: 鬱.[12] This would suggest that depression is a universal phenomenon, yet debate rages in the field of cross-cultural psychiatry, where a postmodern turn moves our perspective beyond the narrow focus on the Western liberal self and onto a comparative, relativistic view of depression. Surely this kind of anthropological view can tell us whether depression is a constant feature of human experience? In fact, there are divergences within the study of depression across different cultures and nations. The most extreme postmodernist view is that depression (and mental disease in general) is entirely local to the society (or period) because it is culturally, not biologically, determined. Local customs set the rules for what is normal and abnormal, healthy and diseased. For the Zuni, native people of Arizona, it can be admirable to be in an apparently severe condition of resignation and passivity, something that an American psychiatrist would regard as Major Depressive Disorder. These rules can apply to historical cultures: the valorisation of melancholy in certain periods of European and American history is similarly culture bound.

Arthur Kleinman—Harvard psychiatrist and the central figure in the anthropology of mental illness for the last forty years or so—has argued that Chinese depressives present their illness as somatic symptoms such as back pain and do not actually feel depressed. Chinese culture mitigates against the expression of personal sadness and loneliness, says Kleinman, whereas Western cultural norms mean that Westerners are more free to show their loneliness and sadness in certain circumstances, although prejudice still persists even in the allegedly enlightened West to this day. The difficulty for Kleinman and others of this postmodern and culture-bound perspective is that it now becomes unclear in what sense depression can meaningfully be said to exist at all. How can lower back pain be equated with sadness? Where is the common pathology? Kleinman is equivocal on the issue because he persists in referring to a phenomenon called 'depression' that is experienced across cultures and even historical eras, even if it might be differently expressed.[13] Others argue that 'in fact, every culture recognises depression in forms that Westerners would recognize' and that the Chinese sample did report 'high rates of DSM-III style symptoms when asked about them'.[14] The issue here seems to have been how Chinese depressives presented themselves socially rather than what they were actually experiencing. Kleinman's research, together with other cross-cultural studies, has been criticised for not taking into account distinctions between normal and pathological sadness, whether culture bound or not.

The fundamental point of disagreement between relativistic cross-cultural studies and other perspectives that emphasise universal aspects of depression, such as biological models, evolutionary psychiatry, or even certain forms of psychodynamic theory, is the problem that one cannot compare illness in different societies unless one has some idea of the universal. To compare depression in China and America, one has to have an underlying definition of depression. Are there certain biological facts that are undeniable and that make it possible to make claims about the wrongness or rightness of judgements about a disorder? For example, some Southerners in antebellum America believed that runaway slaves had a mental disorder: we know this not to be true, and that knowledge helps correct a social injustice. More contentious is Horwitz and Wakefield's claim that depression is a facet of human evolution, and that this is perfectly compatible with the statement that local cultural contexts shape the form that the universal underlying biological processes take. They make the careful but contested distinction between normal sadness, which functions as a useful evolutionary adaptation, and depression, which is a disorder and a dysfunction of the loss–response mechanism.

Final thoughts: back to melancholia?

We conclude with a consideration of the calls for what might be termed a return to melancholy: a demand from theorists and practitioners of various persuasions that we reinstate a model of the human that escapes the reductionism of biochemical definitions. This is not to end where we began, and certainly not to deny possible benefits of biochemical research, but to learn lessons from the long story of what we now call depression. At one end of this call to recognise melancholia, some psychoanalysts advocate a return to the vision of the whole person rather than the biological unit: 'as we read through paper after paper on depression considered as a brain disease, we totally lose any sense that at the core of many people's experience of inertia and lack of interest in life lies the loss of a cherished human relationship or a crisis of personal meaning'.[15] Cognitive Behavioural therapists too, despite their tension with Freudian psychoanalysis, put the thinking and feeling human at the centre of their treatments.

William Styron has called for a return to an old, but perhaps more accurate, way of describing an illness that has a deeper and more profound meaning than the new biomedical definition of depression of the postmodern period can encompass: 'I felt the need ... to register a strong protest against the word "depression."... "Melancholia" would still appear to be a far more apt and evocative word for the blacker forms of the disorder, but it was usurped by a noun with a bland tonality and lacking any magisterial presence, used indifferently to describe an economic decline or a rut in the ground, a true wimp of a word for such a major illness.'[16] Perhaps, as Styron suggested even back in 1990 at the start of the Prozac revolution, returning to the old, rich word 'melancholia' would signal a sea change in the way we define and treat depression. His words are still relevant to the present. This move would not necessarily be a total break from the biochemical model of the Prozac days (whatever the mechanism, the drugs *do* seem to work for some people some of the time), but the new melancholia would be more focused on the social and personal psychology of the individual, rather than merely positing theoretical deficiencies in one's genetic or chemical make-up. Some areas of psychiatry attempt to achieve this aim already, but many fall prey to the exigencies of the particular situation and the pressure to prescribe drugs as quick fixes. How the competing perspectives outlined here will move forward is unclear, and is patently a complicated matter.

The calls for the return of social psychiatry and an acknowledgement of the social pressures on people rather than individual chemical pathologies fit into this new agenda. It has also been argued recently that the stage is set for the return of anxiety and drugs to treat it, and, I also argue, since anxiety has formed part of the melancholic complex since the classical period, this might be a chance to reunite the two in a more helpful formulation than the narrow focus on Major Depressive Disorder that has characterised the DSM-III definition.[17] Edward Shorter has recently and controversially pointed out that the DSM-V draft does not look greatly encouraging, but at least has introduced a category of 'mixed anxiety–depression' that acknowledges the age-old connection of the two, a link evident in the earlier chapters of this book. Otherwise Shorter—equally controversially—finds that the new manual 'fixes none of the problems with the previous DSM series, and even creates some new ones'.[18] The theoretical chaos of the 1970s, the restrictive solution of the DSM-III, and its subsequent reliance on drug treatments have dominated the recent story of depression, yet it is clear that the 'new' biomedical depression of the late twentieth and twenty-first centuries needs to evolve or be remade for a more richly human and specific vision of this protean but very real illness.

By examining the history of melancholia and depression, we can see that the depressed patient is not reducible to a biochemically deficient machine, but an individual embedded in a complex social environment. Treating depression beyond the 'magic bullet' model may prove to be expensive as it will tackle social as well as personal factors, and require more extended talking cures as well as drugs, but such difficulties need to be overcome if we are to move usefully beyond the Prozac age.

GLOSSARY

ACEDIA the depressive languor associated with a monastic existence, originally from Egyptian desert monks in the fourth century, gave rise to a set of symptoms that included a nostalgia for their previous lives and a hatred for the present monastic one, low mood, ennui and general misery.

AFFECTIVE disorder—one of mood, as opposed to **cognitive**, one of thought.

ALIENISTS a nineteenth-century term for 'mad doctors'.

ANIMAL SPIRITS thought of as similar to a superfine liquid, were transmitted via the blood to different parts of the body, including the brain, and acted as mediators between the governing soul and the material body. They could become corrupted by any number of factors and thus affect the whole body, including the brain. They persisted into the eighteenth century, where the **nerves** were seen as vessels (hollow or solid, like musical strings) for the animal spirits. The **fibres**, the solid parts of the body, could also be affected by malfunctions in the spirits and nerves.

ARISTOTELIAN *melancholy* a mode derived from Aristotle's pronouncements on the genius of the melancholic, and roughly opposed to the **Galenic**, which took Galen's more sceptical and pragmatic approach to melancholy as an illness to be suffered rather than celebrated.

ASTHENIA the term in the early nineteenth century 'Brunonian' theory that explained melancholy as a state of under-stimulation or under-excitement. Mania was **sthenic** overstimulation.

BLACK BILE (Gk. *melaina chole*) the melancholy humour and one of the four humours supposed to constitute the body in ancient medicine. Hallucinations or delusions were supposedly caused by **vapours** produced by overheated black bile. It was translated into Latin as *atra bilis* and into English as black bile.

DEMENTIA PRAECOX a nineteenth-century term initially, now known roughly as schizophrenia.

DIAGNOSTIC AND STATISTICAL MANUAL OF MENTAL DISORDERS (DSM) the bible of modern American psychiatry, first published in 1952, and now approaching its fifth version.

EGO (sense of self-identity) and **superego** (conscience or conscious moral self that polices the ego)—both complex terms in Freudian psychoanalysis. The ego tends to be anxious about the influence of the **id** (the uncontrolled primal desires of the animal being) and the superego, as it mediates between the two. Opportunities for depression in this conflictual scenario are rife.

ENDOGENOUS DEPRESSION (of internal causation, within the individual, apparently 'causeless') and **exogenous depression** (externally motivated). These terms have been mapped onto other ways of defining depression and melancholia more or less roughly according to the particular theoretical perspective. So **endogenous** came to refer to a pattern of symptoms deemed to be more serious and/or

psychotic rather than the milder **neurotic** type. The term **neurotic** came to dominate **reactive/exogenous** for the reason that most kinds of depressions seemed to have an environmental stress that predisposed the sufferer to the illness.

ENGLISH MALADY an eighteenth-century name for nervous disorders, of which depression was one.

ETIOLOGY the study of causes of disease.

HUMOURS The traditional four humours composing the body and temperament were akin to different fluids (literally 'sap' or 'juice'). Cure of melancholy in the long-lived humoral theory dating from the Ancient Greeks would consist of attempts to correct the humoral imbalance by purging the excess melancholy humour from the blood (bloodletting, leeching, for example) or by certain drugs such as **hellebore**, effectively a poison that caused immediate diarrhoea and vomiting.

HYPOCHONDRIES the area just below the ribcage—in humoral medicine in which the excess of black bile from digestive disturbance supposedly caused 'an atrabilious evaporation [which] produces melancholic symptoms of mind by ascending to the brain like a sooty substance or a smoky vapour'. The flatulence from this type of melancholy caused it to be dubbed 'windy melancholy' or 'hypochondriacal melancholy' before the twentieth century. Thus the close relationship of the organs allegedly involved in the production of black bile and its regulation became roughly cognate names for the condition of melancholia in the early modern period: **Hypochondria (or the Hypp), Spleen, Hysteria, and Vapours.** After the eighteenth century **Hysteria** (notoriously) became a distinct clinical entity.

IATROCHEMICAL, IATROMECHANICAL from the Greek for doctor (*iatroi*)—both post-humoral ways of approaching the body as a chemical or hydraulic machine.

INVOLUTIONAL MELANCHOLIA a depressive disorder thought to strike in late middle age or old age and to be accompanied by paranoia.

LYPEMANIA Esquirol coined this term in the nineteenth century. For him, melancholy was now a disorder of the emotions, not intellect, and should be called 'lypemania'. It was a form of **monomania**, a partial insanity ('gay or sad') focused on one object.

MAJOR DEPRESSIVE DISORDER (MDD) a contemporary definition of depression. It is a **unipolar** disorder, which is separate from manic depression, a **bipolar** disorder.

MORAL MANAGEMENT non-medical treatment of lunatics in the nineteenth century.

MANIC DEPRESSION nowadays **bipolar disorder**; **unipolar** depressions do not involve mania.

NEURASTHENIA 'exhaustion of the nervous system' and a close cognate of depression in the nineteenth and early twentieth centuries.

NEUROSIS originally referring to physical disorders of the nerves in the eighteenth century but redefined by Freud as stemming from unconscious sexual conflicts. In psychoanalysis a **neurotic depressive** still has contact with reality, but a **psychotic depressive** has loosened or broken the link to reality. Formerly, psychotic depression broadly meant melancholia with delusions.

NON-NATURALS or those six factors described in the Ancients and continuing into the eighteenth century that were considered to be within the control of the individual person, such as aliments (food and drink); the environment (climate, air); sleep and waking; exercise and rest; evacuations (faecal, urinal, sexual and so on); passions and mental state.

PATHOLOGY the study and diagnosis of disease; to be pathological is to be related to or caused by disease.

SELECTIVE SEROTONIN REUPTAKE INHIBITORS (SSRIs) a recent class of antidepressants including Prozac.

SIMPLE a medicinal plant possessing a particular healing quality.

UNITARY PSYCHOSIS the notion that all mental disease moves along a single continuum of severity.

NOTES

Prologue: Dr Samuel Johnson (1709–1784)

1. James Boswell, *Boswell's Life of Johnson: Together with Boswell's Journal of a Tour to the Hebrides, and Johnson's Diary of a Journey into North Wales*, ed., George Birkbeck Hill, revised and enlarged edition L. F. Powell, 6 vols (Oxford: Clarendon Press, 1934), i:483.

2. Ibid. iv:362, 'To Mr Langton, August 25 1784'.

3. Ibid. iv:425.

4. Ibid. i:487.

5. Ibid. i:35.

6. Ibid. iv:427.

7. Ibid. iv:215–16.

8. 'Samuel Johnson's Melancholy', *The Anatomy of Madness* (ed.), Bynum, Porter and Shepherd, 2 vols (London: Tavistock, 1985), i:63–88.

9. Boswell, *Life of Johnson*, i:73, 'Johnson a frolicksome fellow'.

10. Ibid, i:63–4.

11. George Irwin, *Samuel Johnson: a Personality in Conflict* (Auckland: Auckland University Press, 1971), 60.

12. Boswell, *Life of Johnson*, i:65.

13. Ibid. i:64.

14. Samuel Johnson, *The Rambler* no. 85 January 8, 1751, in *The Works of Samuel Johnson*, ed., W. J. Bate and Albrecht B. Strauss, vol. 4, *The Rambler* (New Haven: Yale University Press, 1969), 83.

15. Boswell, *Life of Johnson*, i:298.

16. Ibid. i:277.

17. 'Johnson's Vile Melancholy', in *The Age of Johnson*, ed., F.W. Hilles and W.S. Lewis (New Haven: Yale University Press, 1949), 3–14.

18. E. L. McAdam, Jr, ed. *Samuel Johnson, Diaries, Prayers and Annals* (New Haven: Yale University Press, 1958), 25 April, 1752; 6 May, 1752; 6 May, 1752; Easter 1757; Easter 1753.

19. Boswell, *Life of Johnson*, ii:229.

20. Johnson, *The Rambler*, no. 134, June 29, 1751, in *The Works of Samuel Johnson*, vol. 4, *The Rambler*, 346–7.

21. Samuel Johnson, *The History of Rasselas, Prince of Abissinia*, 2 vols (London : Dodsley, 1759), Vol. 2, chap. XLV, 142.

22. Boswell, *Life of Johnson*, iii:185.

23. Sir John Hawkins, *The Life of Samuel Johnson, LL.D.*, 2nd edn, rev. (London: printed for J. Buckland et al., 1787), 545.

24. Boswell, *Life of Johnson*, i:483.

25. Ibid. i:496.

26. *The Letters of Samuel Johnson*, ed., R.W. Chapman (Oxford: Oxford University Press, 1952), vol. 1, no. 311.1a, 331–2.

27. Boswell, *Life of Johnson*, iii:86–7.

28. Ibid. iii:415.

29. Ibid. ii:382.

30. Ibid. ii:440.

31. Mrs Thrale, in *Johnsonian Miscellanies* (ed.), George Birkbeck Hill, 2 vols (Oxford: Clarendon Press, 1897), i:200.

32. Boswell, *Life of Johnson*, iii:447.

33. Johnson, *The Rambler*, no. 89, January 22, 1751, in *The Works of Samuel Johnson*, vol. 4, *The Rambler*, 346–7.

34. Johnson, *The Idler*, no. 41 January 27, 1759, in *Works of Samuel Johnson*, ed. W. J. Bate, J. M. Bullitt, and L. F. Powell, vol. 2, *The Idler* (New Haven: Yale University Press, 1963), 131.

35. Boswell, *Life of Johnson*, iii:98.

36. Ibid.

37. Ibid.

38. Ibid. i:483.

39. Ibid. iii:99.

40. Hawkins, *Life of Samuel Johnson*, 546.

41. Ibid. 577.

1 'Poor Wretch'

1. Quoted and trans. by Peter Toohey, *Melancholy, Love, and Time: Boundaries of the Self in Ancient Literature* (Ann Arbor: University of Michigan Press, 2004), p. 44; *Argonautica*, 4.1313–18.

2. Quoted in Julia Kristeva, *Black Sun: Depression and Melancholia*, trans. Leon S. Roudiez (New York: Columbia University Press, 1989), 7; *Iliad* VI, 200–3.

3. http://www.louvre.fr/llv/oeuvres/detail_notice.jsp?content<>cnt_id=10134198673225454&current_llv_notice<>cnt_id=10134198673225454&folder<>folder_id=985272369650078&bmLocale=en

4. Hippocrates, *Works of Hippocrates*, trans. and ed. W. Jones and E. Withington, 4 vols (Cambridge, MA: Harvard University Press, 1923–31), iv:185.

5. William Vernon Harris, *Restraining Rage: the ideology of anger control in classical antiquity* (Harvard University Press, 2001), 17. See also Ruth Padel's poetic and idiosyncratic *Whom Gods Destroy*, in which she argues that, in the realm of literature, angry, violent and mostly mad melancholics continue in that vein until the eighteenth century, when they then become depressive in the familiar sense.

6. Galen, *On the Affected Parts*, trans. and ed. Rudolph Siegel (Basel: S. Karger, 1976), 93.

7. *On the Affected Parts*, 92–3.

NOTES TO PP. 30–43

8. Ibid. 90–91.

9. Quoted in Toohey, *Melancholy, Love, and Time*, 86.

10. Wellcome Library, London. Four types of hellebore (Helleborus species): flowering stems and floral segments. Coloured lithograph. Iconographic Collections. Library reference no.: ICV No 45236.

11. Jennifer Radden, ed., *The Nature of Melancholy: from Aristotle to Kristeva* (New York: Oxford University Press, 2000), 57; Book XXX of *Aristotle: Problems*, vol. 16, trans. W.S. Hett (Cambridge, MA: Harvard University Press, 1957), Loeb Classical Library.

12. Quoted in Radden, ed., *The Nature of Melancholy*, 57.

13. Celsus, *De Medicina*, trans. W. Spencer, 3 vols (Cambridge, MA: Harvard University Press, 1953–1961), 299–301.

14. Rufus, *Oeuvres de Rufus D'Ephèse*, ed. and trans. C. Daremberg and C. Ruelle (Paris: J. Baillière, 1879), 455.

15. Ibid. 455.

16. John Cassian, *The Twelve Books…on the Institutes of the Coenobia*, trans. and ed. E. Gibson, in Philip Schaff and Henry Wace, ed., *A Library of the Nicene and Post-Nicene Fathers of the Christian Church*, 2nd ser., 14 vols (Grand Rapids, Michigan: W.B. Erdmans, 1955), xi:266–7.

17. 'The Persones Tale', *The Complete Works of Geoffrey Chaucer*, ed. W. Skeat (Oxford: Clarendon, 1894), vol. 4, 613.

2 Genius and Despair

1. Douglas Trevor, *The Poetics of Melancholy in early modern England* (Cambridge: Cambridge University Press, 2004), 6–7.

2. Robert Burton, *Anatomy of Melancholy*, ed. A. R. Shilleto, 3 vols (London: George Allen & Unwin, 1926–27), i:418–19.

3. John Donne, *Devotions upon Emergent Occasions*, ed. John Sparrow (Cambridge: Cambridge University Press, 1923), 69.

4. Lawrence Babb, *Elizabethan Malady: a Study of Melancholia in English Literature from 1580 to 1642* (East Lansing: Michigan-State University Press, 1951), 66.

5. Hannah Allen, [1683] *A Narrative of God's Gracious Dealings With that Choice Christian Mrs Hannah Allen*, in Allan Ingram, ed. *Voices of Madness: Four Pamphlets, 1683–1796* (Stroud: Sutton Publishing, 1997), 19.

6. Timothy Bright, *A Treatise of Melancholy* (London, 1586), 100. (Wellcome Library, London, Printed text published by W. Stansby, London: 1613, Collection: General Collections).

7. Lemnius Levinus, *The Touchstone of Complexions*, trans. Thomas Newton (London, 1576), fol. 148v.

8. André Du Laurens, *A Discourse of the Preservation of the Sight: Of Melancholike Diseases...*, trans. Richard Surphlet (London, 1599), 96; Burton, *Anatomy*, i.455.

9. F.N. Coeffeteau, *A Table of Humane Passions*, trans. Edward Grimeston (London: Nicholas Okes, 1621), 192–93.

10. Burton, *Anatomy*, i.448–9; Pierre de La Primaudaye, *The French Academie*, trans. T. Bowes (London, 1618), 467.

11. Burton, *Anatomy*, i.441, 475.

12. For this and the following, see Katharine Hodgkin, *Madness in Seventeenth-Century Autobiography* (Basingstoke: Palgrave, 2007), 82, 85.

13. L0015116 Wellcome Library, London 'Melencola I', after Albrecht Dürer. Engraving 1514. By: Albrecht Dürer, Collection: Iconographic Collections.

14. Noel L. Brann, *The Debate Over the origin of Genius during the Italian Renaissance: The theories of supernatural frenzy and Natural Melancholy in Accord and in Conflict on the Threshold of the Scientific Revolution* (Brill: Leiden, 2002), 24.

15. Michele Savonarola, *Practica major* (Venice: Juntas, 1547), tract. VI, cap. 1, rubr. 14, fol. 69r.

16. Pavian physician Giammeteo Ferrari da Grado (d.1472), *Practica seu commentaria in nonum Rasis ad Almansorem* (Venice: Juntas, 1560), cap. 11, fol. 56r.

17. V0007670ER Wellcome Library, London. A depressed scholar surrounded by mythological figures; representing the melancholy temperament. Etching by J.D. Nessenthaler after himself, *c.*1750. By: Johann David Nessenthaler. Published: Johann Michael Probst Aug[usta] V[indicorum] [Augsburg. Collection: Iconographic Collections, Library reference no.: ICV No 7891R.

18. Quoted in Hodgkin, *Madness in Seventeenth-Century Autobiography*, 64.

19. Robert Burton also calls black bile 'the Devil's bath'.

20. St Caterina of Genoa, *Il Dialogo Spirituale*, in *Opus Catharinianum*, 2 vols (Genoa: Marietti, 1962), ii.398.

21. St Antonino, *Summa theological in quatuor partes distributita*, Par I, tit. 9, cap. 2, in *Opera Omnia*, 2 vols (Verona: Apud Augustinum Carattonium, 1740), ii, col. 938.

22. Marsilio Ficino, *De Vita Libri Tres* (Basel, 1549, first pub. 1482–89).

23. Babb, *The Elizabethan Malady*, 65; Ficino, De Vita Libri Tresi 19–20.

24. André Du Laurens, *A Discourse*, 86.

25. Trevor, *The Poetics of Melancholy*, 32.

26. Trevor, *The Poetics of Melancholy*, 20.

27. 'L'Allegro', *Poems of Mr John Milton* (London, 1645), 30, lines 1–7.

28. 'Il Penseroso', p. 37.

29. L0031473 Wellcome Library, London. BURTON, Robert (1577–1640). R. Burton, *The anatomy of melancholy: what it is, with all the kinds, causes, symptomes, prognostickes, & seuerall cures of it, in three partitions, with their severall sections, members,& subsections…By Democritus Jumior* [i.e. Robert Burton]. *With a satyricall preface conducing to the following discourse.* London: Peter Parker, 1676. Frontispiece engraved by C. le Blon. Collection: Rare Books Library reference no.: EPB and EPB.

30. William Shakespeare, *As You Like It*, 4.1.10–18.

31. Andrew Solomon, *The Noonday Demon: An Anatomy of Depression* (London: Vintage, 2002), 305.

32. 'Studious She is and all Alone' frontispiece, *The Philosophical and Physical Opinions*, written by her Excellency, the Lady Marchionesse of Newcastle (London, 1655).

33. Lesel Dawson, *Lovesickness and gender in early modern English literature* (Oxford: Oxford University Press, 2008), 103–7.

34. Katharine Hodgkin, *Madness in Seventeenth-Century Autobiography* chapter 4 'Melancholy: A Land of Darkness', 60–85. Hodgkin uses the work of Carol Thomas Neely to show that a new form of female melancholy arises where women's melancholy is not the genius type but one related to the 'disordered female wombs and genitals' (Neely, cited by Hodgkin, 83).

35. Quoted in Hodgkin, *Madness in Seventeenth-Century Autobiography*, 67; Lambeth Palace Library, MS Sion arc; L40.2/E47, f.7r.

3 From Spleen to Sensibility

1. Reprinted in Clark Lawlor and Akihito Suzuki, eds, *Sciences of Body and Mind*, Vol. 2 of *Literature and Science 1660–1834*, gen. ed. Judith Hawley, 8 vols (London: Pickering and Chatto, 2003), 69.

2. Lawrence Babb, *The Elizabethan Malady; A Study of Melancholia in English Literature from 1580 to 1642* (East Lansing: Michigan-State University Press, 1951), 28.

3. Thomas Willis, *Two Discourses Concerning the Soul of Brutes*, trans. S. Pordage (London, 1683), 188.

4. Friedrich Hoffmann, *A System of the Practice of Medicine*, trans. W. Lewis and A. Duncan, 2 vols (London, 1783), ii:300.

5. Friedrich Hoffmann, *Fundamenta Medicinae*, trans. Lester S. King (London: MacDonald, 1971), 71.

6. Hermann Boerhaave, *Boerhaave's Aphorisms: Concerning the Knowledge and Cure of Diseases* (London: W. and J. Innys, 1735), 312–13.

7. Edward Shorter and David Healy, *Shock Therapy: A History of Electroconvulsive Treatment in Mental Illness* (New Jersey: Rutgers University Press, 2007), 271.

8. *The Medical Works of Richard Mead* (London: C. Hitch, 1762), xxv.

9. Richard Mead, *Medical Precepts and Cautions*, trans. T. Stack (London: J. Brindley, 1751), 79.

10. Henry Fielding, *Amelia*, 4 vols (London: A. Millar, 1752), Vol. 1, Bk III, Chp. VII, 218.

11. Nicholas Robinson, 'of the Hypp',*Gentleman's Magazine*, Vol. 2, November 1732, 1062–4.

12. William Cullen, *First Lines of the Practice of Physic*, new edn, 4 vols (Edinburgh: Elliot and Cadell, 1786), iv:126.

13. William Cullen, *Institutions of Medicine. Part I. Physiology*, 3rd edn, (Edinburgh: Elliot, 1785), 96–101.

14. See Allan Ingram, ed., *Patterns of Madness* (Liverpool: Liverpool University Press, 1998), 83 ff.

15. V0001110 Wellcome Library, London. George Cheyne. Mezzotint by J. Faber, junior, 1732, after: J. van Diest. Collection: Iconographic Collections Library reference no.: ICV No 1297.

16. George Cheyne, *The English Malady*, 33–9, in Ingram, *Patterns of Madness*, 86.

17. William Stukeley, *Of the Spleen* (London, 1723), 25; John F. Sena, *The English Malady: The Idea of Melancholy from 1700 to 1760*, PhD Dissertation, Princeton University (1967), 184.

18. Sir Richard Blackmore, *Essays upon Several subjects*, 2 vols (London, 1717), i:193.

19. Blackmore, 'An Essay upon the Spleen', in *Essays Upon Several Subjects*, ii:258–9.

20. Sena, *The English Malady*, 186.

21. Blackmore, *Essays upon Several Subjects*, i:214.

22. Nicholas Robinson, *A New System of the Spleen, Vapours and Hypo-chondriack Melancholy* (London: A. Bettesworth et al., 1729), 209.

23. John Dryden, *Absalom and Achitophel*, II.163–4.

24. Sena, *The English Malady*, 174.

25. Samuel A. Tissot, *Three Essays: First, On the Disorders of People of Fashion, Second, On Diseases Incidental to Literary and Sedentary Persons, Third, On Onanism: Or, a Treatise upon the Disorders produced by Masturbation: or, the Effects of Secret and Excessive Venery*, trans. Francis Bacon Lee, M. Danes, A. Hume, MD (Dublin: James Williams, 1772).

26. Roy Porter, *Mind Forg'd Manacles: A History of madness in England from Restoration to the Regency* (London: Penguin, 1990), 108–9.

27. Kenneth Dewhurst, *Dr Thomas Sydenham (1624–1689): his Life and Original Writings* (Berkeley and LA: University of California Press, 1966), 174.

28. Bernard Mandeville, *A Treatise of Hypochondriack and Hysterick Passions* (London, 1711, this edn 1730), 177.

29. I am indebted to Heather Meek's essay 'Creative Hysteria and the Intellectual Woman of Feeling', in *Figures et culture de la dépression (1660–1800)/The Representation and Culture of Depression*, the *European Spectator*, vols 10 and 11, ed. Clark Lawlor and Valérie Maffre (Montpellier: Presses universitaires de la Méditerranée, 2011), 87–98.

30. Reprinted in Lawlor and Suzuki, *Sciences of Body and Mind*, 69.

31. *Correspondence of Thomas Gray*, ed. Paget Toynbee and Leonard Whibley (Oxford: Clarendon Press, 1935), i:209, 'Gray to West, May 27, 1742'.

32. Sarah Scott, *The History of Sir George Ellison*, ed. Betty Rizzo (Lexington: University Press of Kentucky, 1996), 35.

33. *A forensic dispute on the legality of enslaving the Africans, held at the public commencement in Cambridge, New-England, July 21st, 1773….* (Boston, 1773).

34. See, for example, Sarah Fielding. *The history of Ophelia. Published by the author of David Simple*, 2 vols (London: R. Baldwin, 1760).

35. (London: J. Roberts, 1721).

36. Cited in George Howe Colt, *The Enigma of Suicide* (New York: Simon and Schuster, 1992), 169.

37. Cited in Elizabeth A. Dolan, 'British Romantic melancholia: Charlotte Smith's Elegiac Sonnets, medical discourse and the problem of sensibility', *Journal of European Studies*, (2003) vol. 33, 237–53, 239; Charlotte Smith, *Conversations Introducing Poetry: Chiefly on Subjects of Natural History. For the Use of Children and Young Persons* (London: J. Johnson, 1804), 100. I am indebted to Pauline Morris's unpublished PhD thesis for this discussion.

4 Victorians, Melancholia, and Neurasthenia

1. J. E. Esquirol, 'Mélancholie', in *Dictionnaire des Sciences Médicales par une Société de Médicins et de Chirurgiennes* (Paris: Panckoucke, 1820), 148; quoted in G. E. Berrios, 'Mood Disorders, Clinical Section', *A History of Clinical Psychiatry*, ed. German E. Berrios and Roy Porter (London: Athlone, 1995), 385–407, 389.

2. See G.E. Berrios, 'Mood Disorders, Clinical Section', *A History of Clinical Psychiatry*, 385–407.

3. Berrios, *A History of Clinical Psychiatry*, 397. L0022595 Wellcome Library, London 'Melancholy passing into mania' by W. Bagg after a photograph by H. W. Diamond. Lithograph. From: *The Medical Times and Gazette*. Published: John Churchill, London, January 2 to June 26, 1858, volume 37, part i, facing page 246, plate 4. Collection: Wellcome Images.

4. Daniel Noble, *Elements of Psychological Medicine: An Introduction to the Practical Study of Insanity* (London: Churchill, 1855), xii.

5. Arthur C. Benson, *Thy Rod and Thy Staff* (London: Smith, Elder, 1912), 1.

6. Janet Oppenheim, *Shattered Nerves: Doctors, Patients and Depression in Victorian England* (Oxford: Oxford University Press, 1991), 84.

7. See L0026686 Wellcome Library, London. A woman diagnosed as suffering from melancholia. Colour lithograph, 1892, after J. Williamson, 1890. 1890–1892. By: J. Williamson after: Byrom Bramwell Published: [s.n.],[Edinburgh]: [1892] Printed: McLagan & Cumming Lith.) (Edin[burg]h).

8. Edwin Lee, *Treatise on Some Nervous Disorders*, 2nd edn (London: Burgess and Hill, 1838), 16.

9. Samuel Tuke, *Description of the Retreat, An Institution Near York for Insane Persons of the Society of Friends* (York: W. Alexander, 1813), 151–2; Allan Ingram, ed., *Patterns of Madness in the Eighteenth Century: A Reader* (Liverpool: Liverpool University Press, 1998), 240.

10. Ingram, *Patterns of Madness*, 239.

11. Philippe Pinel, *A Treatise on Insanity…*, trans. D. D. Davis (Sheffield: W. Todd, 1806), 101.

12. J. E. Esquirol, *Mental Maladies. A Treatise on Insanity*, trans. E. K. Hunt (Philadelphia: 1845), 203, expanded from an essay in 1819.

13. See L0022709 Wellcome Library, London. Monomania and depression. From: Outline of lectures on mental diseases. By: Alexander Morison. Published: London, 1826, facing page 138, plate VII.

14. V0009108ER Wellcome Library, London. A man whose face exemplifies the melancholy temperament. Drawing, *c.*1789. 1789 after: Johann Caspar Lavater and Thomas Holloway. Published: [s.n.], [London]: *c.*1789 Collection: Iconographic Collections Library reference no.: ICV No 9348R.

15. Cited in Oswald Doughty, 'The English Malady of the Eighteenth Century', *The Review of English Studies*, (1926), vol. II, no. 7, 257–69, 268.

16. Johann Christian Heinroth, *Textbook of Disturbances of Mental Life: Or Disturbances of the Soul and Their Treatment*, trans. J. Schmorak and intro. George Mora, 2 vols (Baltimore: Johns Hopkins University Press, 1975 [1818]), 1:189.

17. Heinroth, *Textbook*, 190–1.

18. Ibid. 2:362–3.

19. E. von Feuchtersleben, *The Principles of Medical Psychology*, trans. H. Evans Lloyd and B. G. Babington (London: Sydenham Society, 1847 [1845]).

20. Griesinger, *Mental Pathology and Therapeutics*, 2nd edn, trans. C. Lockhart Robertson and James Rutherford (London: New Sydenham Society, 1867), 1–9.

21. Porter, in Porter and Berrios, ed., *A History of Clinical Psychiatry*, 418–19.

22. J. C. Prichard, *A Treatise on Insanity and other Disorders Affecting the Mind* (London: Sherwood, Gilbert and Piper, 1835), 18.

23. D. Hack Tuke and John Bucknill, *A Manual of Psychological Medicine* (London: John Churchill, 1858), 178.

24. Henry Maudsley, *The Pathology of Mind. A study of its Distempers, Deformities and Disorders* (London: MacMillan, 1895).

25. Richard Von Krafft-Ebing, *Text-Book of Insanity based on Clinical Observations...*, trans. Charles G. Chaddock, intro. Frederick Peterson (Philadelphia: F. A. Davis, 1904), xiii.

26. Charles Mercier, 'Melancholia', in D. Hack Tuke (ed.), *A Dictionary of Psychological Medicine*, 2 vols (Philadelphia: P. Blakiston, 1892), 2:788.

27. George Beard, 'Neurasthenia, or Nervous Exhaustion', *Boston Medical and Surgical Journal*, (1869), N.S. 3, 217–21, 217.

28. Maurice de Fleury, *Medicine and the Mind*, trans. Stacy B. Collins (London: Downey and Co., 1900).

29. V0011623 Wellcome Library, London. A common cold germ asking the father of a neurasthenia bacillus if he can marry her; he is refused on account of the social gap between them. Pen drawing by C. Harrison, 1913. By: Charles Harrison. Published: 1913. Collection: Iconographic Collections Library reference no.: ICV No 11888. Dialogue includes: Proud parent 'You cannot have my daughter the social gulf is too wide remember

you are a mere germ of a common cold—she is a bacillus of neurasthenia.'

30. Thomas Dixon Savill, *Clinical Lectures on Neurasthenia*, 3rd ed., rev. and enl. (London: Longman, Brown, 1858), 188.

31. *The Memoirs of John Addington Symonds*, ed. Phyllis Grosskurth (Chicago: University of Chicago Press, 1986), 64.

5 Modernism, Melancholia, and Depression

1. Kraepelin, *Psychiatrie. Ein Lehrbuch für Studirende und aerzte*, 5th ed. (Leipzig: Johann Ambrosius Barth, 1896), 2:359–61; trans. by Adolf Meyer in Eunice Winters, *The Collected Papers of Adolf Meyer*, 4 vols (Baltimore: The Johns Hopkins Press, 1951) 2:355–6.

2. *Clinical Psychiatry: A Text-Book for students and Physicians*, ed. A. Ross Defendorf, (New York: Macmillan, 1902), 299–300—from 6th German edn of Kraepelin's *Lehrbuch der Psychiatrie*.

3. Quoted in Jennifer Radden, ed., *The Nature of Melancholy: From Aristotle to Kristeva* (Oxford: Oxford University Press, 2000), 261; extract from Kraeplein, 'Manic-depressive Insanity', in the *Textbook of Psychiatry*, 8th edn, 1909–15, trans. Mary Barclay and ed. George Robinson (Edinburgh: E&S Livingstone, 1920).

4. Ross (ed.), *Clinical Psychiatry*, 302.

5. Quoted in Radden, *The Nature of Melancholy*, 265.

6. Ibid. 266.

7. Ibid. 271.

8. Ibid. 277.

9. Karl Abraham, 'Notes on the psychoanalytical investigation and treatment of manic-depressive insanity and allied conditions', in *Selected papers of Karl Abraham, M.D.*, ed. Douglas Bryan and Alix Strachey, trans. (London: Hogarth, 1927 [1911]), 137–56.

10. 'Mourning and Melancholia' (1917 [1915]), *The standard edition of the complete psychological works of Sigmund Freud*, trans. from the German under the general editorship of James Strachey, in

collaboration with Anna Freud, Vol.14 (1914–1916), *On the history of the psycho-analytic movement; Papers on metapsychology and other works* (London: Hogarth and the Institute of Psycho-analysis, 1957), 237–58.

11. Sandor Radó, *Psychoanalysis of Behaviour; Collected Papers*, 2 vols (New York: Grune and Stratton, 1956–1962), i:48.

12. Otto Fenichel, *The Psychoanalytic Theory of Neurosis* (New York: W.W. Norton, 1945), 387.

13. Melanie Klein, 'A contribution to the psychogenesis of manic-depressive states', in *Contributions to psycho-analysis: 1921–1945*, intro. Ernest Jones, (London: Hogarth, 1948 [1935]), 282–310.

14. Edward Bibring, 'Mechanism of Depression', in Phyllis Greenacre, ed., *Affective Disorders: Psychoanalytic Contributions to their Study* (New York: International University Press, 1953), 17.

15. Edith Jacobson, *Depression: Comparative Studies of Normal, Neurotic, and Psychotic Conditions* (New York: International University Press, 1971), 180.

16. Adolf Meyer, *Collected Papers of Adolf Meyer*, ii:142–3.

17. Adolf Meyer, *Psychobiology: A Science of Man*, ed. Eunice Winters and Anna Mae Bowers (Springfield, IL.: Charles Thomas, 1957), 158.

18. Meyer, *Psychobiology*, 174–5.

19. Laura D. Hirshbein, *American Melancholy: constructions of depression in the twentieth century* (New Brunswick: Rutgers University Press, 2009), 15.

6 The New Depression

1. American Psychiatric Association, *Diagnostic and Statistical Manual of Mental Disorders* (Washington D.C.: Author, 1952), 33–4; Allan Horwitz and Jerome C. Wakefield, *The Loss of Sadness: How Psychiatry Transformed Normal Sorrow into Depressive Disorder* (Oxford: Oxford University Press, 2007), 86.

2. DSM-I, 25.

3. American Psychiatric Association, *Diagnostic and Statistical Manual of Mental Disorders*, 2nd ed. (Washington D.C.: Author, 1968), 40.

4. The story of the Feighner Criteria's creation is complicated but well described in Horwitz and Wakefield, *Loss of Sadness*, 91–5.

5. Horwitz and Wakefield, *Loss of Sadness*, 96.

6. Ibid. 103.

7. Ibid. 107.

8. Ibid. 173.

9. See, for example, Martin A. Kohli et al., 'The Neuronal Transporter Gene *SLC6A15* Confers Risk to Major Depression', *Neuron* 70:2, 28 April, 2011, pp. 252–65.

10. See E. S. Valenstein, *Blaming the Brain* (New York, Free Press 1998), 101.

11. Edward Shorter, *Before Prozac: The Troubled History of Mood Disorders in Psychiatry* (New York: Oxford University Press, 2008), 48.

12. *The Paris Review* (Flushing, New York), Winter 1999, vol. 41, Issue 153, p. 118, lines 1–2, 7.

13. Gerald Klerman, quoted in M.C. Smith, *A Social History of the Minor Tranquillizers* (New York: Pharmaceutical Products Press, 1985), 89.

14. David Healy, *Mania: A Short History of Bipolar Disorder* (Baltimore: The Johns Hopkins University Press, 2008), xvi.

15. Shorter, *Before Prozac*, 10, 125.

16. Jim Read, *Psychiatric Drugs: Key Issues and Service User Perspectives* (Basingstoke: Palgrave, 2009).

17. Mikkel Borch-Jacobsen, *Making Minds and Madness: From Hysteria to Depression* (Cambridge: Cambridge University Press, 2009), 197–204.

7 'The Drugs Don't Work'? The Future for Depression and Melancholia

1. Aaron T. Beck, *Cognitive Therapy and Emotional Disorders* (Madison Ct.: International University Press, 1975), 110; Jennifer Radden,

ed., *The Nature of Melancholy: From Aristotle to Kristeva* (Oxford: Oxford University Press, 2000), 322.

2. Quoted in Jennifer Radden, ed., *The Nature of Melancholy*, 313; from Martin Seligman, *Helplessness: On Depression, Development and Death* (New York: W.H. Freeman, 1975).

3. Radden, *The Nature of Melancholy*, 315.

4. Ibid. 315.

5. Ibid. 315.

6. Ibid. 316.

7. See Joel Paris, *The Fall of an Icon: Psychoanalysis and Academic Psychiatry* (Toronto: University of Toronto Press, 2005).

8. Allan Horwitz and Jerome C. Wakefield, *The Loss of Sadness: How Psychiatry Transformed Normal Sorrow into Depressive Disorder* (Oxford: Oxford University Press, 2007), 206.

9. George Brown and Tirril Harris, *Social origins of depression: A study of psychiatric disorder in women* (London: Tavistock, 1978).

10. Jim Read, *Psychiatric Drugs: Key Issues and Service User Perspectives* (Basingstoke: Palgrave, 2009), 63.

11. Oswald Doughty, 'The English Malady of the Eighteenth Century', *The Review of English Studies*, 1926, vol. II, no. 7, 257–69, 267.

12. My thanks to Dr Junko Kitanaka for allowing me to discuss her work on Japanese depression.

13. Jennifer Radden, 'Is This Dame Melancholy?: Equating Today's Depression and Past Melancholia', *Philosophy, Psychiatry, & Psychology*, 10:1, (2003), 37–52, 44.

14. Horwitz and Wakefield, *The Loss of Sadness*, 200.

15. Darian Leader, *The New Black: Mourning, Melancholia and Depression* (London: Penguin, 2009), 20–21.

16. Quoted in Dan G. Blazer, *The Age of Melancholy: 'Major Depression' and its social origins* (New York: Routledge, 2005), 36, originally in William Styron, *Darkness Visible: A Memoir of Madness* (New York: Random House, 1990), 36, 37.

17. Allan Horwitz, 'How an Age of Anxiety Became an Age of Depression', *The Milbank Quarterly*, 88:1, 2010, 112–38.

18. Edward Shorter, 'Why Psychiatry Needs Therapy: A manual's draft reflects how diagnoses have grown foggier, drugs more ineffective', *Wall Street Journal*, 'Life and Culture' section, 27 February 2010.

FURTHER READING

The history of depression and melancholia is long, complicated, and has been fortunate in attracting the attention of a multitude of great scholars. The present history is naturally indebted to this work and, in this particular essay, draws attention to the tip of a very large iceberg. What follows is merely indicative and, inevitably, I will have omitted works on which I have drawn here: I can only apologise in advance.

General

The work of Jennifer Radden in *The Nature of Melancholy: from Aristotle to Kristeva* (Oxford: Oxford University Press, 2000) is a good place to start as it is both a general anthology of some key texts on depression and a discussion of the difficult issues associated with the equation of past melancholia with contemporary depression. Her *Moody Minds Distempered: Essays on Melancholy and Depression* (Oxford: Oxford University Press, 2009) picks up again on many of the general philosophical problems that we encounter throughout this book. Stanley W. Jackson's magisterial *Melancholia and Depression; From Hippocratic Times to Modern Times* (London: Yale University Press, 1986) surveys key thinkers and concepts from classical times up to the early 1980s in the manner of a traditional medical history, but is less sensitive to questions of gender, class, and ethnicity than contemporary scholarship, and is inevitably silent on the developments associated with Prozac and the DSM-III to V. *A History of Clinical Psychiatry*, eds German E. Berrios and Roy Porter (London: Athlone, 1995) remains a good resource into the more technical history of the disease in context with other conditions, and does provide some cultural history into the bargain. Roy Porter, William F. Bynum, and Michael Shepherd have edited the *Anatomy of Madness: Essays in the history of psychiatry*, 3 vols (New York: Routledge, 2003), also a fine mine of general information.

Prologue

Much has been written on Johnson's depression, but the original biographies by his melancholic friend Boswell (*Boswell's Life of Johnson: Together with Boswell's Journal of a Tour to the Hebrides and Johnson's Diary of a Journey into North Wales*, ed. George Birkbeck Hill, 6 vols (Oxford: The Clarendon Press, 1934)) and Sir John Hawkins (*The Life of Samuel Johnson, LL.D.*, 2nd edn, rev. (London: J. Buckland et al., 1787)) provide contrasting and entertaining perspectives. Johnson's own works, including his letters, diaries, and prayers, are fascinating insights into the tortured and melancholic mind of a genius.

Katherine Balderston's 'Johnson's Vile Melancholy' (in *The Age of Johnson*, eds F.W. Hilles and W.S. Lewis (New Haven: Yale University Press, 1949), 3–14) amusingly puts forward the controversial argument about Johnson's alleged sexual peccadilloes, but Roy Porter's interpretation 'Samuel Johnson's Melancholy', (*The Anatomy of Madness*, ed. Bynum, Porter, and Shepherd, 2 vols (London: Tavistock, 1985), i:63–88) argues more convincingly for the influence of religious guilt on Johnson's condition. Walter Jackson Bate's biography—*Samuel Johnson*

(London: Chatto and Windus, 1977)—is the most attentive to Johnson's melancholia, although it takes a more Freudian perspective than is fashionable now. Peter Martin's *Samuel Johnson: A Biography* (London: Weidenfeld & Nicholson, 2008) is an accessible recent summary of scholarly research and deals well with his melancholy.

Chapter 1

Classical melancholia is a foreign entity indeed, and scholars have discussed both the actual definition of melancholia at the time and instances of what we might now consider to be depression. Peter Toohey's *Melancholy, Love, and Time: Boundaries of the Self in Ancient Literature* (Ann Arbor: University of Michigan Press, 2004) is a highly contested version of events, but nevertheless argues that modern depression is to be found in classical culture and its art. Toohey's representational evidence is slender, however, and certainly not proven or unchallenged: see Shadi Bartsch, Review of Toohey, *Melancholy, Love, and Time*, in *The Classical Review* vol. 55 no. 2, 498–99. Toohey is also criticised in William Vernon Harris's *Restraining Rage: the ideology of anger control in classical antiquity* (Harvard University Press, 2001), 17. Ruth Padel's *Whom Gods Destroy* is a classic study that depicts melancholia as angry rather than sad until the shift in the eighteenth century. Debra Hershkowitz's *The Madness of Epic: Reading Insanity from Homer to Statius* (Oxford: Clarendon Press, 1998) discusses tensions between (religious) literary and (secular) Hippocratic depictions of melancholia. The lack of relationship between classical mania and modern manic depression is explained by David Healy's *Mania: A Short History of Bipolar Disorder* (Baltimore: The Johns Hopkins University Press,

2008)—mania might emerge from severe melancholia in the classical conception.

All students of classical and Renaissance melancholy must turn to Raymond Klibansky, Erwin Panofsky, and Fritz Saxl, *Saturn and Melancholy: Studies in the history of natural philosophy, religion, and art* (New York: Basic Books, 1964), which describes the pseudo-Aristotelian tradition of melancholic genius. Noga Arikha's more recent book on the history of the humours, *Passions and Tempers: A History of the Humours* (New York, NY: Ecco, 2007), is helpful in this regard, and for later periods up to the present day, as well as more sensitive to gender difference.

The bridge from the classical period to the Renaissance requires consideration of acedia, and Siegfried Wenzel's *The Sin of Sloth: Acedia in Medieval thought and literature* (Chapel Hill: University of North Carolina Press, 1967) shows how the monastic disease evolved throughout the period. Stanley W. Jackson discusses the technical aspects of this disease in 'Acedia the Sin and Its Relationship to Sorrow and Melancholia', in Arthur Kleinman and Byron Good (eds), *Culture and Depression: Studies in the Anthropology and Cross-cultural Psychiatry of Affect and Disorder* (London: University of California Press, 1985), also a volume worth examining for its overall theme.

Chapter 2

The link of melancholy and genius in its Renaissance form is explored by Noel L. Brann's *The Debate Over the origin of Genius during the Italian Renaissance: The theories of supernatural frenzy and Natural Melancholy in Accord and in Conflict on the Threshold of the*

Scientific Revolution (Brill: Leiden, 2002), Winfried Schleiner, *Melancholy, Genius, and Utopia in the Renaissance* (Wiesbaden: Harrassowitz, 1991) and Erwin Panofsky's *Albrecht Dürer* (Princeton: Princeton University Press, 1945). Noga Arikha is helpful here, as she is on the position of women in Renassiance melancholy. In the specifically British context, the great work of Robert Burton is key, and the scholarly edition is *The Anatomy of Melancholy*, Thomas C. Faulkner, Nicolas K. Kiessling, and Rhonda L. Blair (eds) (Oxford: Clarendon Press, [1621] 1989). Useful work that recasts Burton and the function of melancholy in Renaissance culture includes Angus Gowland, *The Worlds of Renaissance melancholy: Robert Burton in context* (Cambridge: Cambridge University Press, 2006) and his compendious article 'The Problem of Early Modern Melancholy', *Past & Present*, 2006, 191(1): 77–120. Jeremy Schmidt's *Melancholy and the Care of the Soul: Religion, Moral Philosophy and Madness in Early Modern England* (Aldershot: Ashgate, 2007) again follows the recent trend of seeing melancholy as the construction of a set of competing discourses (medical, legal, nationalistic, religious) rather than merely as a medical condition to be equated with contemporary depression.

Melancholy is a literary disease in this period, and useful older studies of its presence in literature include Lawrence Babb's *The Elizabethan Malady. A Study of Melancholia in English Literature from 1580 to 1642* (East Lansing: Michigan State College Press, 1951) and Bridget Gellert Lyons, *Voices of melancholy; studies in literary treatments of melancholy in Renaissance England* (New York: Barnes & Noble, 1971). More recently, Douglas Trevor's *The poetics of melancholy in early modern England* (Cambridge: Cambridge University Press, 2004) brings Hamlet into a more complex dialogue with recent critical trends, one clear difference being the role of gender. The problematic role of women in melancholy is tackled in a less than convincing way in Juliana Schiesari's *The gendering of melancholia: feminism, psychoanalysis, and the symbolics of loss in Renaissance literature* (New York: Cornell University Press, 1992). Two more historically accurate interventions in the realm of gender are Lesel Dawson's *Lovesickness and gender in early modern English literature* (Oxford: Oxford University Press, 2008) and Katharine Hodgkin's *Madness in Seventeenth-Century Autobiography* (Basingstoke: Palgrave, 2007). For a powerful autobiographical account of religiously inflected early modern melancholia, see Hannah Allen, *A Narrative of God's Gracious Dealings With that Choice Christian Mrs. Hannah Allen* [1683], in Allan Ingram (ed.) *Voices of Madness: Four Pamphlets, 1683–1796* (Stroud: Sutton Publishing, 1997).

Michael MacDonald's *Mystical Bedlam: Madness, Anxiety, and Healing in Seventeenth-Century England* (Cambridge: Cambridge University Press, 1981) remains a classic, and, on the issue of suicide, Michael MacDonald and Terence R. Murphy's *Sleepless Souls: Suicide in early modern England* (Oxford: Clarendon Press, 1990) argues that suicide came to be seen as a medical problem during the eighteenth century rather than a moral and religious sin. Rab Houston challenges this view in his recent *Punishing the Dead: Suicide, Lordship and community in Britain 1500–1830* (Oxford: Oxford University Press, 2011). *The History of Suicide in England, 1650–1850*, ed. by Mark Robson, Paul S. Seaver, Kelly McGuire, Jeffrey Merrick, and Daryl Lee (London: Pickering and Chatto, 2011–12) presents a window into the variety of 'long' eighteenth-century suicide.

Chapter 3

The eighteenth century's supposed secularisation has been heavily debated in terms of how melancholy is framed, and some core studies take on this issue: John F. Sena, *The English Malady: The Idea of Melancholy from 1700 to 1760*, PhD Dissertation, Princeton University (1967) and Roy Porter, *Mind Forg'd Manacles: A History of madness in England from Restoration to the Regency* (London: Penguin, 1990). Many of Porter's very readable works deal with melancholy. Marvelously scholarly, if dated, studies of literary representations are Amy Louise Reed's *The Background of Gray's Elegy: A Study in the Taste for Melancholy Poetry, 1700–51* (New York: Columbia University Press, 1924) and, pressing forward into the Romantic period, Eleanor M. Sickels's *The Gloomy Egoist: Moods and Themes of Melancholy from Gray to Keats* (New York: Columbia University Press, 1932). Useful works (even if I do say so myself) representing a variety of perspectives on melancholia and its cognates arising from the *Before Depression* project at Northumbria University can be found at http://www.beforedepression.com and include Allan Ingram, Stuart Sim, Clark Lawlor, Richard Terry, John Baker, and Leigh Wetherall Dickson, *Melancholy Experience in Literature of the Long Eighteenth Century: Before Depression, 1660–1800* (Basingstoke: Palgrave Macmillan 2011) and *Figures et culture de la dépression (1660–1800)/The Representation and Culture of Depression*, the *European Spectator*, vols 10 and 11, eds Clark Lawlor and Valérie Maffre (Montpellier: Presses universitaires de la Méditerranée, 2011). Again, the position of women, class, nation, and ethnicity is more fully analysed than in earlier classic studies of the period like Sena's.

Chapter 4

No bibliography of nineteenth-century depression would be complete without Janet Oppenheim's excellent *Shattered Nerves: Doctors, Patients and Depression in Victorian England* (Oxford: Oxford University Press, 1991). For a more institutional slant see Joseph Melling and Bill Forsythe, *The Politics of Madness: The State, Insanity and Society in England, 1845–1914* (London: Routledge, 2006), while Elaine Showalter's *The Female Malady* (New York: Pantheon, 1987) remains a key feminist intervention. Oppenheim deals well with neurasthenia, but see also Tom Lutz's *American Nervousness, 1903: An Anecdotal History* (New York: Cornell University Press, 1991) and Marijke Gijswijt-Hofstra and Roy Porter, eds, *Cultures of Neurasthenia: From Beard to the First World War* (Clio Medica 63) (Amsterdam: Rodopi, 2001). For literary representations of neurasthenia, see Kate Chopin's *The Awakening* (New York, NY: Bantam Classic, 1981,[1899]), Frank Norris's *The Pit* (Doubleday, Page & Co. 1903), Edith Wharton's *House of Mirth* (New York, NY: C. Scribner's sons, 1905), Jack London's *Martin Eden* (New York, NY: Macmillan Co., 1909) and Theodore Dreiser's *The 'Genius'* (New York, NY: John Lane, 1915).

Chapter 5

The modern historiography of twentieth-century depression has been well served by Laura D. Hirshbein's *American Melancholy: constructions of depression in the twentieth century* (New Brunswick: Rutgers University Press, 2009), which describes the role of popular culture and the active power of the consumer in generating the version of depression we have with us today. Hirshbein's work is exemplary of a move beyond the stories of 'great men' heroically discovering cures: hers is a much more nuanced vision of all the (gendered) participants in the social construction of a disease. She also makes it clear that the way depression is represented in cultural media is important in its ongoing definition and treatment. Dan G. Blazer's The *Age of Melancholy: 'Major Depression' and its social origins* (New York: Routledge, 2005) makes a plea for the return of social psychiatry, and attempts to attribute the rise of depression in recent years to factors beyond the chemical or genetic; postmodern hopelessness has replaced the positive thrust of modernism according to this perspective. Junko Kitanaka's excellent study of *Depression in Japan: Psychiatric Cures for a Society in Distress* (Princeton: Princeton University Press, 2011) provides a cross-cultural perspective which again stresses the need to acknowledge and treat the social factors involved in depression. Postmodernism in the round is represented by Mikkel Borch-Jacobsen's *Making Minds and Madness: From Hysteria to Depression* (Cambridge: Cambridge University Press, 2009). The 'construction' or 'invention' of depression in recent times is also, and with a greater historical specificity, discussed in Gary Greenberg's *Manufacturing Depression: The Secret History of a Modern Disease* (London: Bloomsbury, 2010).

Chapter 6

Allan Horwitz and Jerome C. Wakefield's *The Loss of Sadness: How Psychiatry Transformed Normal Sorrow into Depressive Disorder* (Oxford: Oxford University Press, 2007) makes the powerful argument for a redefinition of depression into a much more narrow entity, thus making room for normal sadness, which at the moment is massively pathologised. This book covers the formation of the DSM-III and much besides, and is essential reading for anyone interested in how the history of depression matters to its present: history has material effects on the actual treatment of actual patients.

The deeply contentious subject of treatment, whether it be drugs or other somatic remedies, is bracingly engaged by the publications (too many to list here) of the twin scourges of mainstream psychiatry and psychoanalysis, Edward Shorter and David Healy. Their *Shock Therapy: A History of Electroconvulsive Treatment in Mental Illness* (New Jersey: Rutgers University Press, 2007) is a good place to start for the myths and realities of ECT. Shorter's *Before Prozac: The Troubled History of Mood Disorders in Psychiatry* (New York: Oxford University Press, 2008), is representative of his assault on both institutional regulation of drugs as well as 'Big Pharma' in the allegedly inappropriate treatment of depressed patients. David Healy's *The Antidepressant Era* (London: Harvard University Press, 1999) and *Let Them Eat Prozac: the unhealthy relationship between the pharmaceutical industry and depression* (New York: New York University Press, 2004) are excoriating attacks that do what the subtitle of the latter suggests. Shorter and Healy are practising psychiatrists, and are not averse to the appropriate use of drugs or biomedical thinking, as Shorter and Max Fink's *Endocrine Psychiatry: Solving the Riddle of Melancholia* (New York: Oxford University Press, 2010) demonstrates.

Chapter 7

Other attacks on mainstream psychiatry and the drug industry in particular are legion, and from a variety of perspectives: two of note are Joanna Moncrieff's *The Myth of the Chemical Cure:*

A *Critique of Psychiatric Drug Treatment* (Basingstoke: Palgrave, 2007) and Richard Bentall's *Doctoring the Mind: Why Psychiatric Treatments Fail* (London: Allen Lane, 2009). Fighting the corner for psychoanalysis is Darian Leader's *The New Black: Mourning, Melancholia and Depression* (London: Penguin, 2009), while Julia Kristeva's *Black Sun: Depression and Melancholia,* trans. L.S. Roudiez (New York: Columbia University Press, 1989) provides an older and more post-structuralist-feminist version of the Freudian tradition.

The effectiveness of psychiatric drugs from a user perspective is sensibly discussed by Jim Read's publication for Mind (the mental health charity)—*Psychiatric Drugs: Key Issues and Service User Perspectives* (Basingstoke: Palgrave, 2009). A scholarly contribution that proposes action on the coalface of primary care from the stable of Cambridge historian and psychiatrist German Berrios is Christopher Callahan and German Berrios's *Reinventing Depression: A History of the Treatment of Depression in Primary Care, 1940–2004* (New York: Oxford University Press, 2005).

The depression memoir thrives in our age, as it has in others, and good examples are easy to find: three elegant books by male sufferers and writers are William Styron's *Darkness Visible: A Memoir of Madness* (New York: Random House, 1990), Andrew Solomon's *The Noonday Demon: An Anatomy of Depression* (London: Vintage, 2002), and Lewis Wolpert's *Malignant Sadness: The Anatomy of Depression,* 3rd edn, (London: Faber and Faber, 2006). The most famous narrative of all is Elizabeth Wurtzel's *Prozac Nation: Young and Depressed in America* (New York: Houghton and Mifflin, 1994). Journalist Stephanie Merritt's more recent *The Devil within: A Memoir of Depression* (London: Vermilion, 2009) movingly describes her postnatal depression. The optimistic 'recovery' genre of depression writing is exemplified by sufferer-turned-mental nurse Irene Burnett-Thomas's *Turning the blues around: one woman's story of kicking depression to nurse the mentally ill* (London: Athena Press, 2009). The future of depression lies as much with a recognition of the patient's voice and experience as with advances in biomedical technology, so these texts should not be regarded as a literary extra to the 'real' business of depression research.

BIBLIOGRAPHY

Abraham, K., 'Notes on the psychoanalytical investigation and treatment of manic-depressive insanity and allied conditions', in *Selected papers of Karl Abraham, M.D.*, ed. Douglas Bryan and Alix Strachey, trans. (London: Hogarth, 1927 [1911]).

Allen, H., *A Narrative of God's Gracious Dealings With that Choice Christian Mrs Hannah Allen*, in Allan Ingram, ed. *Voices of Madness: Four Pamphlets, 1683–1796* (Stroud: Sutton Publishing, 1997).

American Psychiatric Association, *Diagnostic and Statistical Manual of Mental Disorders* (Washington D.C.: Author, 1952).

——*Diagnostic and Statistical Manual of Mental Disorders*, 2nd ed. (Washington D.C.: Author, 1968).

Arikha, N., *Passions and Tempers: A History of the Humours* (New York, NY: Ecco, 2007).

Aristotle, Book XXX of *Aristotle: Problems*, vol. 16, trans. W. S. Hett (Cambridge, MA: Harvard University Press, 1957).

Babb, L., *Elizabethan Malady: a Study of Melancholia in English Literature from 1580 to 1642* (East Lansing: Michigan-State University Press, 1951).

Bartsch, S., 'Review of Toohey, *Melancholy, Love, and Time*', in *The Classical Review* vol. 55 no. 2, 498–99.

Bate, W. J., and Strauss, A. B., *The Works of Samuel Johnson*, 4 vols (New Haven: Yale University Press, 1969).

Bate, W. J., *Samuel Johnson* (London: Chatto and Windus, 1977).

Beard, G., 'Neurasthenia, or Nervous Exhaustion', *Boston Medical and Surgical Journal*, (1869), N.S. 3, 217–21.

Beck, A. T., *Cognitive Therapy and Emotional Disorders* (Madison Ct.: International University Press, 1975).

Benson, A. C., *Thy Rod and Thy Staff* (London: Smith, Elder, 1912).

Bentall, R., *Doctoring the Mind: Why Psychiatric Treatments Fail* (London: Allen Lane, 2009).

Berrios, G. E., 'Mood Disorders, Clinical Section', *A History of Clinical Psychiatry*, ed. German E. Berrios and Roy Porter (London: Athlone, 1995).

Bibring, E., 'Mechanism of Depression', in Phyllis Greenacre, ed., *Affective Disorders: Psychoanalytic Contributions to their Study* (New York: International University Press, 1953).

Blackmore, R., *Essays upon Several subjects*, 2 vols (London, 1717).

Blazer, D. G., The *Age of Melancholy: 'Major Depression' and its social origins* (New York: Routledge, 2005).

Boerhaave, H., *Boerhaave's Aphorisms: Concerning the Knowledge and Cure of Diseases* (London: W. and J. Innys, 1735).

Borch-Jacobsen, M., *Making Minds and Madness: From Hysteria to Depression* (Cambridge: Cambridge University Press, 2009).

Boswell, J., *Boswell's Life of Johnson: Together with Boswell's Journal of a Tour to the Hebrides, and Johnson's Diary of a Journey into North Wales*, ed. George Birkbeck Hill, revised and enlarged edition L. F. Powell, 6 vols (Oxford: Clarendon Press, 1934).

Brann, N. L., *The Debate Over the origin of Genius during the Italian Renaissance: The theories of supernatural frenzy and Natural Melancholy in Accord and in Conflict on the Threshold of the Scientific Revolution* (Brill: Leiden, 2002).

Bright, T., *A Treatise of Melancholy* (London: W. Stansby, 1613).

Brown, G., and Harris, T., *Social origins of depression: A study of psychiatric disorder in women* (London: Tavistock, 1978).

Burnett-Thomas, I., *Turning the blues around: one woman's story of kicking depression to nurse the mentally ill* (London: Athena Press, 2009).

Burton, R., *Anatomy of Melancholy*, ed. A. R. Shilleto, 3 vols (London: George Allen & Unwin, 1926–27).

——*The Anatomy of Melancholy*, Thomas C. Faulkner, Nicolas K. Kiessling, and Rhonda L. Blair (eds) (Oxford: Clarendon Press, [1621] 1989).

Bynum, W. F., Porter, R., and Shepherd, M., eds, *The Anatomy of Madness*, 2 vols (London: Tavistock, 1985).

Callahan, C., and Berrios, G., *Reinventing Depression: A History of the Treatment of Depression in Primary Care, 1940–2004* (New York: Oxford University Press, 2005).

Cassian, J., *The Twelve Books...on the Institutes of the Coenobia*, trans. and ed. E. Gibson, in Philip Schaff and Henry Wace, ed., *A Library of the Nicene and Post-Nicene Fathers of the Christian Church*, 2nd ser., 14 vols (Grand Rapids, Michigan: W. B. Erdmans, 1955).

Celsus, *De Medicina*, trans. W Spencer, 3 vols (Cambridge, MA: Harvard University Press, 1953–1961).

Chapman, R. W., ed., *The Letters of Samuel Johnson* (Oxford: Oxford University Press, 1952.

Chopin, K., *The Awakening* (New York, NY: Bantam Classic, 1981, [1899]).

Coeffeteau, F., N., *A Table of Humane Passions With their Causes and Effects*, trans. Edward Grimeston (London: Nicholas Okes, 1621).

Colt, G. H., *The Enigma of Suicide* (New York: Simon and Schuster, 1992).

Cullen, W., *Institutions of Medicine. Part I. Physiology*, 3rd edn, (Edinburgh: Elliot, 1785).

—— *First Lines of the Practice of Physic*, new edn, 4 vols (Edinburgh: Elliot and Cadell, 1786).

Dawson, L., *Lovesickness and gender in early modern English literature* (Oxford: Oxford University Press, 2008).

de Fleury, M., *Medicine and the Mind*, trans. Stacy B. Collins (London: Downey and Co., 1900).

Dewhurst, K., *Dr Thomas Sydenham (1624–1689): his Life and Original Writings* (Berkeley and L.A.: University of California Press, 1966).

Dolan, E. A., 'British Romantic melancholia: Charlotte Smith's Elegiac Sonnets, medical discourse and the problem of sensibility', *Journal of European Studies*, (2003) vol. 33, 237–53.

Donne, J., *Devotions upon Emergent Occasions*, ed. John Sparrow (Cambridge: Cambridge University Press, 1923).

Doughty, O., 'The English Malady of the Eighteenth Century', *The Review of English Studies*, (1926), vol. II, no. 7, 257–69.

Dreiser, T., *The 'Genius'* (New York, NY: John Lane, 1915).

Du Laurens, A., *A Discourse of the Preservation of the Sight: Of Melancholike Diseases...*, trans. Richard Surphlet (London: F. Kingston, 1599).

Esquirol, J. E., 'Mélancholie', in *Dictionnaire des Sciences Médicales par une Société de Médicins et de Chirurgiennes* (Paris: Panckoucke, 1820).

—— *Mental Maladies. A Treatise on Insanity*, trans. E. K. Hunt (Philadelphia: 1845).

Fenichel, O., *The Psychoanalytic Theory of Neurosis*, (New York: W.W. Norton, 1945).

Ferrari da Grado, G., *Practica seu commentaria in nonum Rasis ad Almansorem* (Venice: Juntas, 1560), cap. 11, fol. 56r.

Fielding, H., *Amelia*, 4 vols (London: A. Millar, 1752).

Fielding, S., *The history of Ophelia. Published by the author of David Simple*, 2 vols (London: R. Baldwin, 1760).

Ficino, M., *De Vita Libri Tres* (Basel: 1549, first pub. 1482–89).

Freud, S., *The standard edition of the complete psychological works of Sigmund Freud*, trans. from the German under the general editorship of James Strachey, in collaboration with Anna Freud, Vol.14 (1914–1916), *On the history of the psychoanalytic movement; Papers on metapsychology and other works* (London: Hogarth and the Institute of Psycho-analysis, 1957).

Galen, *On the Affected Parts*, trans. and ed. Rudolph Siegel (Basel: S. Karger, 1976).

Gijswijt-Hofstra, M., and Porter, R., eds, *Cultures of Neurasthenia: From Beard to the First World War* (Clio Medica 63) (Amsterdam: Rodopi, 2001).

Gowland, A., *The Worlds of Renaissance melancholy: Robert Burton in context* (Cambridge: Cambridge University Press, 2006).

—— 'The Problem of Early Modern Melancholy', *Past & Present*, 2006, 191(1): 77–120.

Gray, T., *Correspondence of Thomas Gray*, ed. Paget Toynbee and Leonard Whibley (Oxford: Clarendon Press, 1935).

Greenberg, G., *Manufacturing Depression: The Secret History of a Modern Disease* (London: Bloomsbury, 2010).

Griesinger, W., *Mental Pathology and Therapeutics*, trans. C. Lockhart Robertson and James Rutherford (2nd edn, London: New Sydenham Society, 1867).

Hack Tuke, D., and Bucknill, J., *A Manual of Psychological Medicine* (London: John Churchill, 1858).

Harris, W. V., *Restraining Rage: the ideology of anger control in classical antiquity* (Cambridge, MA: Harvard University Press, 2001).

Hawkins, J., *The Life of Samuel Johnson, LL.D.* (2nd edn, rev., London: printed for J. Buckland et al., 1787).

Healy, D., *The Antidepressant Era* (London: Harvard University Press, 1999).

——*Let Them Eat Prozac: the unhealthy relationship between the pharmaceutical industry and depression* (New York: New York University Press, 2004).

——*Mania: A Short History of Bipolar Disorder* (Baltimore: The Johns Hopkins University Press, 2008).

Heinroth, J. C., *Textbook of Disturbances of Mental Life: Or Disturbances of the Soul and Their Treatment*, trans. J. Schmorak and intro. George Mora, 2 vols (Baltimore: Johns Hopkins University Press, 1975 [1818]).

Hershkowitz, D., *The Madness of Epic: Reading Insanity from Homer to Statius* (Oxford: Clarendon Press, 1998).

Hill, G. B., *Johnsonian Miscellanies* (Oxford: Clarendon Press, 1897).

Hilles, F. W., and Lewis, W. S., eds, *The Age of Johnson* (New Haven: Yale University Press, 1949).

Hippocrates, *Works of Hippocrates,* trans. and ed. W. Jones and E. Withington, 4 vols (Cambridge, MA: Harvard University Press, 1923–31).

Hirshbein, L. D., *American Melancholy: constructions of depression in the twentieth century* (New Brunswick: Rutgers University Press, 2009).

Hodgkin, K., *Madness in Seventeenth-Century Autobiography* (Basingstoke: Palgrave, 2007).

Hoffmann, F., *A System of the Practice of Medicine*, trans. W. Lewis and A. Duncan, 2 vols (London, 1783).

Horwitz, A., 'How an Age of Anxiety Became an Age of Depression', *The Milbank Quarterly*, 88:1, 2010, 112–38.

—— and Wakefield, J.C., *The Loss of Sadness: How Psychiatry Transformed Normal Sorrow into Depressive Disorder* (Oxford: Oxford University Press, 2007).

Houston, R., *Punishing the Dead: Suicide, Lordship and community in Britain 1500–1830* (Oxford: Oxford University Press, 2011).

Ingram, A., ed., *Patterns of Madness in the Eighteenth Century: A Reader* (Liverpool: Liverpool University Press, 1998).

—— et al., *Melancholy Experience in Literature of the Long Eighteenth Century: Before Depression, 1660–1800* (Basingstoke: Palgrave Macmillan 2011).

Irwin, G., *Samuel Johnson: a Personality in Conflict* (Auckland: Auckland University Press, 1971).

Jackson, S. W., 'Acedia the Sin and Its Relationship to Sorrow and Melancholia', in Arthur Kleinman and Byron Good (eds), *Culture and Depression: Studies in the Anthropology and Cross-cultural Psychiatry of Affect and Disorder* (London: University of California Press, 1985).

—— *Melancholia and Depression; From Hippocratic Times to Modern Times* (London: Yale University Press, 1986).

Jacobson, E., *Depression: Comparative Studies of Normal, Neurotic, and Psychotic Conditions*, (New York: International University Press, 1971).

Johnson, S., *The History of Rasselas, Prince of Abissinia*, 2 vols (London: Dodsley, 1759).

Kitanaka, J., *Depression in Japan: Psychiatric Cures for a Society in Distress* (Princeton: Princeton University Press, 2011).

Klein, M., 'A contribution to the psychogenesis of manic-depressive states', in *Contributions to psycho-analysis: 1921–1945*, intro. Ernest Jones, (London: Hogarth, 1948 [1935]).

Klibansky, R., Panofsky, E., and Saxl, F., *Saturn and Melancholy: Studies in the history of natural philosophy, religion, and art* (New York: Basic Books, 1964).

Kohli, M. A., et al., 'The Neuronal Transporter Gene *SLC6A15* Confers Risk to Major Depression', *Neuron* 70:2, 28 April, 2011, pp. 252–65.

Kraepelin, E., *Psychiatrie. Ein Lehrbuch für Studirende und aerzte*, (5th ed. , Leipzig: Johann Ambrosius Barth, 1896) trans. by Adolf Meyer in Eunice Winters, *The Collected Papers of Adolf Meyer*, 4 vols (Baltimore: The Johns Hopkins Press, 1951).

—— 'Manic-depressive Insanity', in the *Textbook of Psychiatry*, 8th edn, 1909–15, trans. Mary Barclay and ed. George Robinson (Edinburgh: E&S Livingstone, 1920).

Kristeva, J., *Black Sun: Depression and Melancholia*, trans. Leon S. Roudiez (New York: Columbia University Press, 1989).

Cavendish, Margaret, Lady Marchionesse of Newcastle, *The Philosophical and Physical Opinions*, (London, 1655).

Lawlor, C., and Suzuki, A., eds, *Sciences of Body and Mind*, Vol. 2 of *Literature and Science 1660–1834*, gen. ed. Judith Hawley, 8 vols (London: Pickering and Chatto, 2003).

—— and Maffre, V., eds, *Figures et culture de la dépression (1660–1800)/The Representation and Culture of Depression*, the *European Spectator*, vols 10 and 11 (Montpellier: Presses universitaires de la Méditerranée, 2011).

Leader, D., *The New Black: Mourning, Melancholia and Depression* (London: Penguin, 2009).

Lee, E., *Treatise on Some Nervous Disorders*, 2nd edn (London: Burgess and Hill, 1838).

Levinus, L., *The Touchstone of Complexions*, trans. Thomas Newton (London, 1576).

London, J., *Martin Eden* (New York, NY: Macmillan Co., 1909).

Lutz, T., *American Nervousness, 1903: An Anecdotal History* (New York: Cornell University Press, 1991).

Lyons, B. G., *Voices of melancholy; studies in literary treatments of melancholy in Renaissance England* (New York: Barnes & Noble, 1971).

McAdam, Jr, E. L., ed. *Samuel Johnson, Diaries, Prayers and Annals* (New Haven: Yale University Press, 1958).

MacDonald, M., *Mystical Bedlam: Madness, Anxiety, and Healing in Seventeenth-Century England* (Cambridge: Cambridge University Press, 1981).

——and Murphy, T. R., *Sleepless Souls: Suicide in early modern England* (Oxford: Clarendon Press, 1990).

Mandeville, B., *A Treatise of Hypochondriack and Hysterick Passions* (London, 1711, this edn 1730).

Martin, P., *Samuel Johnson: A Biography* (London: Weidenfeld & Nicholson, 2008).

Maudsley, H., *The Pathology of Mind. A study of its Distempers, Deformities and Disorders* (London: MacMillan, 1895).

Mead, R., *Medical Precepts and Cautions*, trans. T. Stack (London: J. Brindley, 1751).

——*The Medical Works of Richard Mead* (London: C. Hitch, 1762).

Meek, H., 'Creative Hysteria and the Intellectual Woman of Feeling', in *Figures et culture de la dépression (1660–1800)/ The Representation and Culture of Depression*, the European Spectator, vols 10 and 11, ed. Clark Lawlor and Valérie Maffre (Montpellier: Presses universitaires de la Méditerranée, 2011).

Melling, J., and Forsythe, W., *The Politics of Madness: The State, Insanity and Society in England, 1845–1914* (London: Routledge, 2006).

Mercier, C., 'Melancholia', in D. Hack Tuke (ed.), *A Dictionary of Psychological Medicine*, 2 vols (Philadelphia: P. Blakiston, 1892).

Merritt, S., *The Devil within: A Memoir of Depression* (London: Vermilion, 2009).

Meyer, A., in Eunice Winters, *The Collected Papers of Adolf Meyer*, 4 vols (Baltimore: The Johns Hopkins Press, 1951).

——*Psychobiology: A Science of Man*, ed. Eunice Winters and Anna Mae Bowers (Springfield, IL.: Charles Thomas, 1957).

Midriff, J., *Observations on the Spleen and Vapours: Containing remarkable cases of person of both sexes and all ranks from the aspiring directors to the humble bubbler who have been miserably afflicted with these melancholy disorders since the fall of the South Sea and other public stocks* (London: J. Roberts, 1721).

Milton, J., *Poems of Mr John Milton* (London: Ruth Raworth for H. Moseley, 1645).

Moncrieff, J., *The Myth of the Chemical Cure: A Critique of Psychiatric Drug Treatment* (Basingstoke: Palgrave, 2007).

Noble, D., *Elements of Psychological Medicine: An Introduction to the Practical Study of Insanity* (London: Churchill, 1855).

Norris, F., *The Pit* (New York: Doubleday, Page & Co., 1903).

Oppenheim, J., *Shattered Nerves: Doctors, Patients and Depression in Victorian England* (Oxford: Oxford University Press, 1991).

Padel, R., *Whom Gods Destroy: Elements of Greek and Tragic Madness*, (Princeton, N.J.: Princeton University Press, 1995).

Panofsky, E., *Albrecht Dürer* (Princeton: Princeton University Press, 1945).

Paris, J., *The Fall of an Icon: Psychoanalysis and Academic Psychiatry* (Toronto: University of Toronto Press, 2005).

Pinel, P., *A Treatise on Insanity…*, trans. D. D. Davis (Sheffield: W. Todd, 1806).

Porter, R., *Mind Forg'd Manacles: A History of madness in England from Restoration to the Regency* (London: Penguin, 1990).

Prichard, J. C., *Treatise on Insanity and other Disorders Affecting the Mind* (London: Sherwood, Gilbert and Piper, 1835).

Primaudaye, P. de la, *The French Academie*, trans. T. Bowes (London, 1618).

Radden, J., ed., *The Nature of Melancholy: from Aristotle to Kristeva* (New York: Oxford University Press, 2000).

—— 'Is This Dame Melancholy?: Equating Today's Depression and Past Melancholia', *Philosophy, Psychiatry, & Psychology*, 10:1, (2003), 37–52.

—— *Moody Minds Distempered: Essays on Melancholy and Depression* (Oxford: Oxford University Press, 2009).

Read, J., *Psychiatric Drugs: Key Issues and Service User Perspectives* (Basingstoke: Palgrave, 2009).

Reed, A. L., *The Background of Gray's Elegy: A Study in the Taste for Melancholy Poetry, 1700–51* (New York: Columbia University Press, 1924).

Radó, S., *Psychoanalysis of Behaviour; Collected Papers*, 2 vols (New York: Grune and Stratton, 1956–1962).

Robinson, N., *A New System of the Spleen, Vapours and Hypochondriack Melancholy* (London: A. Bettesworth et al., 1729).

—— 'of the Hypp', *Gentleman's Magazine*, Vol. 2, November 1732, 1062–4.

Robson, M., et al. eds, *The History of Suicide in England, 1650–1850* (London: Pickering and Chatto, 2011–12).

Ross Defendorf, A., ed., *Clinical Psychiatry: A Text-Book for students and Physicians* (New York: Macmillan, 1902).

Rufus, *Oeuvres de Rufus D'Ephèse*, ed. and trans. C. Daremberg and C. Ruelle (Paris: J. Baillière, 1879).

St Antonino, *Summa theological in quatuor partes distributita*, Par I, tit. 9, cap. 2, in *Opera Omnia*, 2 vols (Verona: Apud Augustinum Carattonium, 1740).

St Caterina of Genoa, *Il Dialogo Spirituale*, in *Opus Catharinianum*, 2 vols (Genoa: Marietti, 1962).

Savill, T., D., *Clinical Lectures on Neurasthenia*, 3rd ed., rev. and enl. (London: Longman, Brown, 1858).

Savonarola, M., *Practica major* (Venice: Juntas, 1547), tract. VI, cap. 1, rubr. 14, fol. 69r.

Sena, J. F., *The English Malady: The Idea of Melancholy from 1700 to 1760*, PhD Dissertation, Princeton University (1967).

Schiesari, J., *The gendering of melancholia: feminism, psychoanalysis, and the symbolics of loss in Renaissance literature* (New York: Cornell University Press, 1992).

Schleiner, W., *Melancholy, Genius, and Utopia in the Renaissance* (Wiesbaden: Harrassowitz, 1991).

Schmidt, J., *Melancholy and the Care of the Soul: Religion, Moral Philosophy and Madness in Early Modern England* (Aldershot: Ashgate, 2007).

Scott, S., *The History of Sir George Ellison*, ed. Betty Rizzo (Lexington: University Press of Kentucky, 1996).

Seligman, M., *Helplessness: On Depression, Development and Death* (New York: W.H. Freeman, 1975).

Shorter, E., *Before Prozac: The Troubled History of Mood Disorders in Psychiatry* (New York: Oxford University Press, 2008).

——'Why Psychiatry Needs Therapy: A manual's draft reflects how diagnoses have grown foggier, drugs more ineffective', Wall Street Journal, 'Life and Culture' section, 27 February 2010.

——and Fink, F., *Endocrine Psychiatry: Solving the Riddle of Melancholia* (New York: Oxford University Press, 2010).

——and Healy, D., *Shock Therapy: A History of Electroconvulsive Treatment in Mental Illness* (New Jersey: Rutgers University Press, 2007).

Showalter, E., *The Female Malady* (New York: Pantheon, 1987).

Sickels, E. M., *The Gloomy Egoist: Moods and Themes of Melancholy from Gray to Keats* (New York: Columbia University Press, 1932).

Skeat, W. ed., *The Complete Works of Geoffrey Chaucer* (Oxford: Clarendon, 1894).

Smith, C., *Conversations Introducing Poetry: Chiefly on Subjects of Natural History. For the Use of Children and Young Persons* (London: J. Johnson, 1804).

Smith, M. C., *A Social History of the Minor Tranquillizers* (New York: Pharmaceutical Products Press, 1985).

Solomon, A., *The Noonday Demon: An Anatomy of Depression* (London: Vintage, 2002).

Stukeley, W., *Of the Spleen* (London, 1723).

Styron, W., *Darkness Visible: A Memoir of Madness* (New York: Random House, 1990).

Symonds, J. A., *The Memoirs of John Addington Symonds*, ed. Phyllis Grosskurth (Chicago: University of Chicago Press, 1986).

Tissot, S. A., *Three Essays: First, On the Disorders of People of Fashion, Second, On Diseases Incidental to Literary and Sedentary Persons, Third, On Onanism: Or, a Treatise upon the Disorders produced by Masturbation: or, the Effects of Secret and Excessive Venery*, trans. Francis Bacon Lee, M. Danes, A Hume, MD (Dublin: James Williams, 1772).

Toohey, P., *Melancholy, Love, and Time: Boundaries of the Self in Ancient Literature* (Ann Arbor: University of Michigan Press, 2004).

Trevor, D., *The Poetics of Melancholy in early modern England* (Cambridge: Cambridge University Press, 2004).

Tuke, S., *Description of the Retreat, An Institution Near York for Insane Persons of the Society of Friends* (York: W. Alexander, 1813).

Valenstein, E. S., *Blaming the Brain* (New York: Free Press, 1998).

von Feuchtersleben, E., *The Principles of Medical Psychology*, trans. H. Evans Lloyd and B. G. Babington (London: Sydenham Society, 1847 [1845]).

Von Krafft-Ebing, R., *Text-Book of Insanity based on Clinical Observations...*, trans. Charles G. Chaddock, intro. Frederick Peterson (Philadelphia: F. A. Davis, 1904).

Wallace, R., 'On Prozac', *The Paris Review* (Flushing, New York), Winter 1999, vol. 41, Issue 153, p. 118.

Wenzel, S., *The Sin of Sloth: Acedia in Medieval thought and literature* (Chapel Hill: University of North Carolina Press, 1967).

Wharton, E., *House of Mirth* (New York, NY: C. Scribner's sons, 1905).

Willis, T., *Two Discourses Concerning the Soul of Brutes*, trans. S. Pordage (London, 1683).

Wolpert, L., *Malignant Sadness: The Anatomy of Depression*, (3rd edn, London: Faber and Faber, 2006).

Wurtzel, E., *Prozac Nation: Young and Depressed in America* (New York: Houghton and Mifflin, 1994).

INDEX